Michael D. Fowler
Sound Worlds of Japanese Gardens

CM00925151

Cultural and Media Studies

Michael D. Fowler is an alumni of the Alexander von Humboldt Foundation's Fellowship program. His research focuses on the intersection between architectural theory, soundscape studies and landscape architecture.

Michael D. Fowler

Sound Worlds of Japanese Gardens

An Interdisciplinary Approach to Spatial Thinking

[transcript]

Bibliographic information published by the Deutsche Nationalbibliothek
The Deutsche Nationalbibliothek lists this publication in the Deutsche Nationalbibliografie; detailed bibliographic data are available in the Internet at http://dnb.d-nb.de

© 2014 transcript Verlag, Bielefeld

All rights reserved. No part of this book may be reprinted or reproduced or utilized in any form or by any electronic, mechanical, or other means, now known or hereafter invented, including photocopying and recording, or in any information storage or retrieval system, without permission in writing from the publisher.

Cover concept: Kordula Röckenhaus, Bielefeld
Cover illustration: The garden of Kyu Furukawa Teien, Tokyo Japan. Copyright Michael Fowler, 2007.
Printed by Majuskel Medienproduktion GmbH, Wetzlar
Print-ISBN 978-3-8376-2568-4
PDF-ISBN 978-3-8394-2568-8

Contents

Foreword

This book is the result of an extended period of research and creative work that has consumed me during the last eight years. It may be initially difficult to categorize a definitive focus discipline that the investigations revealed within this book are wholly aligned to: methods for analysis taken from semiotics, music theory, acoustics and computer science are paired with architectonic form generation strategies, data visualization methods and electro-acoustic sound spatialization techniques. The investigations partake then in what may be dryly referred to as both "scientific" and "artistic" methodologies, though in my opinion, such labels tend to hamper and constrain what the end goal of the approaches used in this book are—that of enabling new ways of what I nominate as an approach to *spatial thinking*. The book then is as much a document of various types of research methods used to address particular questions on the nature of sound design in the Japanese garden, as it is an invitation towards recognizing the potential of the nexus between the scientific and artistic as a rich area for interdisciplinary work. The Japanese garden is of course an important and central feature of this book, but rather than only providing a purely historical account of Japanese garden design or garden aesthetics, the garden is positioned as a spatial exemplar for which a number of interrogations of this exemplar are carried out using a diverse series of critical, conceptual and artistic frameworks. Thus the goal of this book is to enable the analysis, project work and ways of thinking through the proclivities of Japanese garden design to be both standalone investigations as well as, when examined as a whole, representative of the larger framing of *spatial thinking* as an enabling methodology for critical and artistic inquiry.

The first chapter deals with the Japanese garden as an historical object by tracing an arc that covers its early appearance in Japan as essentially a Chinese copy through to its development into the early twentieth-century. By positioning the garden within a larger aesthetic context of the arts of Japan, particular qualities and design techniques are investigated as they relate to the deeper function and context that Japanese gardens arise from. There is of course much extant scholarship regarding the history of the Japanese garden and Japanese art aesthetics. Rather than attempting to recapture what is already within the discourse, I use current research and knowledge as a way to frame what this book seeks to do differently, that of examining the under-investigated nature of the auditory qualities of Japanese garden design as an important element in creating particularly balanced experiences for visitors.

This first chapter then, as an introduction to Japanese garden aesthetics, serves as a basis for the rest of the book, though I use the second chapter as a means to explore the direct relationship that the American composer John Cage had to the Japanese garden. I've been deeply involved in research into Cage's compositions, performance practice, and musical aesthetics, though explore in chapter two one particular graphic music composition called *Ryoanji*—a musical work based on a Japanese garden. I examine *Ryoanji* in relation to Cage's reading and experiences of the famous Zen garden of the same name in Kyoto as well as highlight how a common ground can be found between Cage's aesthetics and those forces and influences at play within the garden design at the temple. Cage's musical work is particularly important for the later investigations undertaken within the book because it represents a type of model or paradigm. Cage essentially translated the garden into a musical score, and it is this strategy and conceptualization of the Japanese garden—as an object whose internal structural relationships might be expressed in various media—that I take as an invitation for developing my own creative works.

But in order to initiate such strategies for what Majorie Siegel would nominate as the *transmediation* of the Japanese garden into other media, the third chapter is firstly concerned with utilizing various methodologies that may assist in revealing those deeper traits regarding the auditory qualities and acoustic properties of Japanese garden design. As such, I introduce various approaches to understanding auditory encounters within Japanese garden design by using, for example, the lens and methodological approach of the discipline of acoustic ecology, the visualization technique of the Greimas semiotic square, and the mathematical constructs of Formal Concept Analysis. Some of the work that forms the basis of this chapter, I have explored previously,[1] though it is synthesized here to form part of a many-layered interrogative strategy for generating spatial information from the Japanese garden. The goal then of the chapter is the collection and analysis of data in order that it may both reveal deeper spatial proclivities, which can in turn point to the nature of Japanese garden design as a multi-sensory approach to composition, while additionally providing suitable spatial information for usurpations in creative and artistic transmediations in the following chapter.

The fourth chapter then is an exploration of the possibilities of taking the analysis, database of raw spatial information, and insights of the previous chapter for use in a number of artistic projects. This chapter introduces and explores a range of multi-channel sound installation projects that are co-paired to a number of architec-

---

1 See in particular: "Hearing a shakkei: the semiotics of the audible in a Japanese stroll garden," *Semiotica* 197 (October 2013): 101-117, "The taxonomy of a Japanese stroll garden: an ontological investigation using Formal Concept Analysis," *Axiomathes* 23/1 (March 2013): 43-59, "Towards an urban soundscaping," in Nathalie Bredella and Chris Dähne, eds., *Infrastrukturen des Urbanen: Soundscapes, Netscapes, Landscapes im kinematischen Raum*, 35-57 (Bielefeld: Transcript, 2013).

tonic modeling projects through the common usurpation of particular Japanese gardens and the analysis of them carried out in the previous chapter. As such, this chapter seeks to continue Cage's *Ryoanji* precedent by attempting to extend, manipulate and ultimately transform the approach and perceived constraints of Cage's original musical framework and musical context. Some of this chapter is also based on previous work,[2] though is bought to light in this chapter as a fundamental element of exploring how the Japanese garden might be re-represented or re-tooled for adaptations in other media. While this chapter seeks not to embrace technical or mathematical means purely for the disposal of the analytical methods themselves, but rather as an aid to delving deeper into the abstract relationship between emergent spatial and artistic forms and the Japanese garden as a spatial database, it does partake somewhat in the language of mathematics as a means to objectively describe the approaches being used. In particular, I outline my use of Voronoi tessellation methods for multichannel electro-acoustic sound spatialization strategies, musical set theory for pitch extraction from soundscape recordings of Japanese gardens, NURBS (Non-uniform Rational B-Spline) modeling strategies for architectonic form finding, Cellular Automatons for controlling digital sound engines and rule-based algorithms used to generated virtual garden spaces. But the use of mathematical abstractions and formulaic descriptions are intended as a foil, and indeed as an aid, to the artistic project work of the chapter which is primarily centered on questions of how might the spatial auditory qualities of Japanese garden design be re-represented in other forms and through other *visual languages*.

As a reflection on the entire methodology presented in the book, I present in the final chapter the deeper context, catalysts and impetus that initiated the work discussed throughout. The analysis, tactics and products of the book are presented then as the methodologies use to enact the idea of *spatial thinking*. But as a means to further contextualize the framework of the notion of *spatial thinking* as an approach that seeks to forge a closer or more intimate nexus between theory and praxis, I also introduce a number of framing terms and concepts. These terms are drawn from the ancient Greek notions of *technê* (the art of making) and *epistemê* (the theoretical knowledge of making). I thus contextualize the work presented in the book as an approach that seeks to synthesize both analytic (or scientific) and creative (artistic) adaptations of spatial information from the Japanese garden. I frame the products and methodologies of *spatial thinking* then as necessarily interdisciplinary inquiries that have sought to blur the boundaries between the formal and the creative and the scientific and the artistic through various research methods, conceptions and attitudes.

---

2 "Sound, aurality and critical listening: disruptions at the boundaries of architecture." *Architecture and Culture* 1/1 (November 2013): 159-178, "Appropriating an architectural design tool for musical ends," *Digital Creativity* 22/4 (January 2012): 275-287, "Transmediating a Japanese garden through spatial sound design," *Leonardo Music Journal* 21 (December 2011): 41-49, "Mapping sound-space: the Japanese garden as auditory model," *Architectural Research Quarterly* 14/1 (March 2010): 63-70, "Shin-Ryoanji: a digital garden," *Soundscape, The journal of Acoustic Ecology* 8/1 (Fall/Winter 2008): 33-36.

In this sense, the Japanese garden represents perhaps both an object of fluidity and a document of an unchanging history. Of course as a catalyst for artistic expression it remains an extremely valuable one that Cage and many other Japanese and Western artists, architects and designers have also recognized and identified. But what is perhaps the Japanese garden's greatest strength is the fact that it can be both identified as an historical object fixed in its spatial predilections as well as a fluid object that is malleable enough to enable adaptations, re-representations and transmediations to flow from its center ever outwards across centuries, countries, cultures and artforms.

Acknowledgments

I would like to firstly thank my wife Gabriele for her support and patience during the writing of this book. This book would not have been possible without the generous support of the Alexander von Humboldt Foundation, The Australian Research Council, Australian Academy of the Humanities, Australia Council for the Arts, Asialink, Australian Network for Art and Technology, Hakodate Future University, Akiyoshidai International Art Village, RMIT University School of Architecture and Design and the Spatial Information Architecture Laboratory. I would also like to thank the Audio Communication Research Group at Technische Universität Berlin and the Japanese Consulate General in Melbourne Australia for their assistance. Many thanks too to the following people who have been inspirational in one way or another: Stuart Gerber, Mark Burry, Laurent Stalder, Leon van Schaik, Stefan Weinzierl, Colin Spiers, Tristram Williams, Chris Dähne, Nathalie Bredella and Sabine von Fischer.

Japanese Garden Aesthetics

An Introduction

> If the bouquets, the objects, the trees, the faces, the gardens, and the texts—if the things and manners of Japan seem diminutive to us [. . .] this is not by reason of their size, it is because every object every gesture, even the most free, the most mobile, seems framed.
> ROLAND BARTHES, *EMPIRE OF SIGNS*, 42.

THE GARDEN AS SPATIAL PARADIGM

Traditional Japanese gardens have been of a profound influence in contemporary garden design throughout modern Japan, as well as a common reference point for Western architectural design aesthetics[1]. They have been recognized as an embodiment of a number of cultural predilections, mapped out and created as a physical manifestation of form through a highly ordered and considered, yet completely natural looking collection of landscape elements and architectonic features. As such, they are universally prized for their unique incorporation of sound events, textures, asymmetric geometry, and the methods to which they involve the visitor in the exploration of a site as a utopian landscape. But having existed in Japan since at least the fifth century CE they are also highly diverse in terms of form, size and function, and range from small dry landscapes of raked gravel with a few features rocks located within the confines of Buddhist temples, large expansive pond gardens originally used for boating and courtly entertainment of the aristocracy, to intimate and unadorned tea gardens used to generate complementary atmospherics for *chanoyu* or the Japanese tea ceremony. But in spite of their large diversity across sites and eras, they are, as Livio Sacchi explains:

1 Andrew Wilson, *Influential Gardeners: The Designers who Shaped 20th-Century Garden Style* (London: Mitchell Beazley/Octopus Publishing, 2002).

"The expression of *mono aware*, man in tune with nature, and at the same time the supreme representation of nature itself. [The Japanese garden] must have no trace of artificial elements. This naturally, makes it extremely artificial, but it doesn't matter, the garden is a representation of an abstract idealized nature, a sort of original earthly paradise or pre-figuration of heavenly paradise."[2]

Notions regarding the Japanese garden as untouched, or as an origin of the natural world, are further emphasized through the notion that a garden should contain the essence of *sabi*, or that that brings a rustic, yearning, "numbness, chilliness [or] obscurity"[3] to the viewer. For larger stroll type gardens, this is balanced with the nature of water to bring sounds and life to the garden through ornamental *koi* (carp), turtles and birds that inhabit lakes and ponds, often in seasonal fluctuations of activity. The creation of distant miniaturized landscapes through the use of dwarfed pines, sculptured shrubs and blue-grey foliage for background plantings (mirroring the human perception of the color of distant scenery) is often combined with foreground treatments. The masterly use of rock placements signifying dry stream beds, shorelines, inlets and beaches, symbolic Buddhist trios and famous mountain peaks of China and Japan are also combined with more abstract compositions that suggest islands that take the form of turtles, cranes or boats. Navigation through a Japanese garden then is always a sensory feast given that plantings are never homogenous but purposely mix deciduous and evergreen varieties to emphasize seasonal cycles of colors, forms, smells and sounds. Paths are created as enticements to explore a garden, bending, curving and using the local topography to conceal vistas until the last moment. Added to this are the types of ground cover and the sounds of different materials underfoot as well as the numerous bridges of stone and wood that signify crossings not only of immediate landscape features of small streams, lakes or ravines, but equally as movements through time and other distant epochs. Thus *mono aware* is a central element in the concept of the Japanese garden in that the goal is to provide what Gunilla Lindberg-Wada describes as "the deeply moving impact on the sensitive, refined beholder of an aesthetically and emotionally moving scene or phenomenon."[4] This notion perhaps reaches a pinnacle with the dry landscape gardens or *kare-sansui* of the Japanese Medieval period in which subtly and austerity are evident in the simplicity and emptiness of the garden. These gardens, though resistive towards specifically representing a natural scene, at the same time embody the aesthetics of nature through an abstract composition of natural materials and seemingly perfect geometries.

---

2 Livio Sacchi, *Tokyo, City and Architecture* (Torino: Skira Editore, 2004), 119.
3 Mikiso Hane, *Modern Japan, A Historical Survey* (Boulder, Colorado: Westview Press, 2001), 21.
4 Gunilla Lindberg-Wada, "Japanese Literary History: The Beginnings," in *Literary History: Towards a Global Perspective, Vol. 1: Notions of Literature across Times and Cultures* (Berlin: Walter de Gruyter GmbH), 132.

As highly influential and imaginative design spaces then, Japanese gardens have also been important catalysts for a diverse series of spatial art practices outside of landscape architecture. But it is perhaps due to the rapid increase of the construction of Japanese gardens outside of Japan since the Second World War[5] that has increased their visibility globally, and thus provided a number of Western artists and designers in the twentieth and twenty-first centuries a glimpse into traditional Japanese aesthetics and Japanese attitudes towards landscape, nature and the relationship between humans and their environment.[6] As a *spatial model* or design exemplar then, the Japanese garden's use of natural materials, site topography, controlled variations of texture and number, and expert manipulation of the viewer's perceived presence, have led numerous explorations into the process of translating these qualities so as to similarly re-produce or curate multi-sensory experiences within a range of other media. These artistic explorations have been varied and range from the highly specific musical composition *Ryoanji* by American composer John Cage (of which I will discuss in depth in the following chapter) or the Koshino House by Tadao Ando,[7] to more general approaches and appropriations of Japanese gardening aesthetics by architect Yoshio Taniguchi,[8] sculptor Isamu Noguchi,[9] and landscape architect Mirei Shigemori.[10] But to firstly trace the source of Japanese garden aesthetics via a more broad discussion of the influences from China and Buddhism on art and landscape gardening will provide a valuable context for understanding some of the techniques and aesthetic goals of Japanese garden design in all its forms and why these have been so valued and recognized by numerous contemporary artists and thinkers.

Early Chinese Models

The introduction to the Japanese archipelago of the religion of Buddhism occurred during the Asuka period (537-621) and is marked by its adoption through the ruling class, principally through the regent Prince Umayado (572-622) who was also known as *Shotoku*. This significant event was the initial impetus for the development

5 Greg Missingham, "Japan 10±, China 1: A First Attempt at Explaining the Numerical Discrepancy between Japanese-style Gardens Outside Japan and Chinese-style Gardens Outside China," *Landscape Research* 32 (2007): 117-146.

6 Gintaras Stauskis, "Japanese Gardens outside of Japan: From the Export of Art to the Art of Export," *Town Planning and Architecture* 35/3 (2011): 212-221.

7 Richard Weston, "Koshino House," in *Plans, Sections and Elevations* (London: Laurence King Publishing, 2004).

8 Yoshio Taniguchi, *The Architecture of Yoshio Taniguchi* (New York: Harry N. Abrams, 1999).

9 Ana Maria Torres, *Isamu Noguchi: A Study of Space* (New York: The Monacelli Press, 2000).

10 Christian Tschumi, *Mirei Shigemori: Modernizing the Japanese Garden* (Berkeley, CA: Stone Bridge Press, 2005).

of a distinct series of ongoing interpretations of Buddhism that found manifestations through numerous works of Japanese art as well as through philosophical discourses and practices. From these initial beginnings, Buddhist art and other related creative practices in Japan eventually wrestled themselves from Chinese models to become a product of native Japanese aesthetics. Buddhism initially arrived in Japan via the Korean peninsula through the ruler of the State of Paekche, King Syöng Myöng. The Korean King sent manuscripts and images of the Buddha in 552 and advised the Japanese Court to adopt the new religion that had already found itself transported from India, through China and Korea. Initial reception and dissemination of Buddhism in Japan was marred by factional disputes and rivalries with the already established religion of Shinto, though these difficulties were eventually overcome. The original documents from Syöng Myöng included *zuzo*, a form of iconographic depiction, that "by the twelfth century [...] were being classified by monks, which led to the standardization of many images"[11] such that the *zuzo* exerted a profound influence on later Japanese Buddhist art through painting, garden making and sculpture. It is notable that these first efforts to introduce and capture the imagination of Japanese society to the edicts of Buddhism were primarily through iconic imagery and art. Indeed George Sansom describes this first encounter with Buddhism as "remarkable that it was those beautiful things rather than to sermons or scriptures that the Japanese people owed their first direct knowledge of the culture which they were about to adopt."[12]

Some of the first examples of new Buddhist art emerging in Japan are the lacquered panel paintings at the Tamamushi Shrine in Nara. These panels, known as *jutaka*, show a precedent for a style that would become popular in later years. The *jutaka* style of painting usually depicted one of the Buddha's previous lives. The fine examples at the Tamamushi Shrine are an embodiment of the "sympathetic and sensitive reception of cultural and technical ideas from earlier Continental [Chinese] Buddhist art."[13] In fact, this first encounter with an influence outside the country's border was assimilated with great speed: "tremendous cultural receptivity in the Asuka period, therefore, may reflect a new Japanese sense of cultural inferiority in the face of an urbane, continental culture and a concomitant yearning for international parity."[14] There is a strong indication then that early Japanese artisans were very conscious of taking such models and re-interpreting them to suit their tastes: "they did not completely detach themselves from T'ang [Chinese] Art, but rather took the decorative elements they had inherited and transformed them into subtle works displaying an elegance that conformed to the [contemporary Japanese] lifestyles and sensibilities."[15]

---

11 Miyeko Murase, *Bridge of Dreams, The Mary Griggs Burke Collection of Japanese Art* (New York: The Metropolitan Museum of Art, 2000), 36.
12 George Sansom, *A History of Japan to 1334* (Tokyo: Charles E. Tuttle, 1963), 64.
13 Sansom, *A History of Japan*, 74.
14 Christine M. E. Guth, *Art, Tea, Industry, Masuda Takashi and the Mitsui Circle* (Princeton: Princeton University Press, 1993), 58.
15 Paul Varley, *Japanese Culture*, 4th ed. (Hawaii: University of Hawaii Press, 2000), 129.

Within Japanese garden design though, those landscapes created for the aristocracy of the Asuka period can only be surmised from archaeological evidence or from references in the 720 *Nihon shoki* (Chronicles of Japan) given that none of the period's gardens have survived through until modern times. But what is apparent from reconstructions and the scant remaining evidence is that the first prototype of the Japanese garden was, much like that of other early Japanese plastic arts and ink paintings, based on Chinese T'ang dynasty models. To this end, the early Japanese garden prototype occupied a particularly large space compared with later styles. The gardens were connected to Palaces of nobility as well as temples within the estates of the court and were commonly populated with large lakes for boating parties as well as miniaturized hillsides, numerous islands, winding streams and rock arrangements in and around the water features. Many of the pleasure gardens also had numerous references to ocean landscapes such as beaches or estuaries and were commonly located at the southern courtyards of the estates.

But within the other plastic arts and painting, reactions towards early Chinese artistic models were initially didactic. Gradually though, aesthetic responses began to curb in what was seen by the Japanese as the unemotional rigidity of Chinese art. Ninth century ash glazed pots from the Shosoin repository display contemporary Japanese artistic attitudes to Chinese models. At Shosoin, earthenware examples are usually well formed by the standard of contemporary T'ang canons, though they are glazed in a particularly non-Chinese fashion. Here, ash is left to fall naturally over the work, creating an uneven green glaze across the shoulders of the pots. Famous schools of ceramic production, particularly the kilns at Bizen, Shigaraki, and Iga, developed controlled chance effects by the variation in use of different clays and kiln temperatures. At Iga, "rugged, asymmetrical flower vases and waste water containers [. . .] in which the warping or cracking that often occurred in the process of firing these heavy-bodied vessels made them all the more desirable."[16] A craftsman in contemporary China might have seen such a work as unusable or disfigured, but the Japanese taste was one in which imperfections, and those creations attributed to the hand of chance, held the most unpredictable, and therefore greatest beauty.

Similarly, the textures of such wares were venerated for their rough, muted quality, their flaws not seen as such but more as characteristics of austerity. Paul Varley comments that: "aesthetically, this was a significant transition, because it represented a reassertion of such basic values as naturalness and irregularity"[17]—two particular features of Japanese gardening aesthetics that become highly prized and firmly established by the beginning of the Nara Era (710-794). Thus, a new aesthetic emerged that was quickly accepted and internalized as a uniquely Japanese approach. In what Fokke Sierksma describes as the second phase in cultural acculturation, a shift begins in which a foreign influence is gradually re-interpreted such that:

16 Guth, *Art, Tea, Industry*, 56.
17 Varley, *Japanese Culture*, 129.

"Objects and ideas are taken over from the strange culture, but derive their meaning from the context of the old culture within which they are now placed. Or again, indigenous elements of culture are given new meaning in the context of the strange new culture."[18]

Thus the ceramics that were created at this time at the kilns of Shigaraki, Iga, and Bizen were in stark contrast to the Chinese models of traditional proportion and absolutism. These new modes of creation from Japanese artists "drew attention to themselves, not for their formal symmetry and decorative refinement, but rather for their irregular forms, softly textured surfaces, and surprising lightness."[19] These ceramic vessels provide early examples of how Japanese culture, after coming into contact with other societies, has absorbed influences and reinterpreted them according to their own aesthetic persuasions.

Figure 1: Byōdō-in, Uji

By the close of the Fujiwara period (897-1185), the establishment of a Japanese style had become particularly evident in painting and in landscape gardening. A shift had occurred regarding Buddhist practices with a new movement called Pure Land Buddhism. The temple and its gardens had begun to find a more central place within the aristocratic court. Whereas during the Asuka period the temple occupied a subsidiary

18 Fokke Sierksma, *Tibet's Terrifying Deities* (Tokyo: Charles E. Tuttle, 1966), 90.
19 Joan Stanley-Barker, *Japanese Art* (London: Thames and Hudson, 1984), 76.

or backdrop to the court, it had now become elevated to assume a more central part of the activities of the court as well as guiding the planning of the site and the integration of the architecture. Pure Land temple gardens represented a paradise that was said to lay to the west where Amida Buddha ruled. One surviving example of a Paradise Garden is located at Byōdō-in in Uji, near Kyoto (Figure 1). It was originally the villa of Fujiwara Michinaga, (966-1028) though his son was responsible in 1053 for the transformation of the villa into a temple that includes the still standing Phoenix Hall. Using the traditional style common in Chinese Song Dynasty temple architecture, though now aligned along an east-west axis, the hall faces a pond containing white stones that represent Mt. Horai (the home of the Eight Immortals of the Daoists'). The Hall is connected to the pond by a small bridge that symbolized the way to paradise. As a meditation and contemplation garden it represents an important prototype for future Japanese gardens in which Buddhist philosophy became an important conceptual tool for the design of the garden.

But these developments and shifts towards a new Japanese aesthetics were also apparent in contemporary painting techniques of the period:

"The growing apart of the two cultures [Japan and China] can be seen in a comparison of painting techniques. In Northern Song [CE 960-1126], the flat, colorist tradition of T'ang painting was replaced by a linear modeling technique called *cunfua*, where depth and texture are defined in brush strokes rather than shading. This eventually gave way to an ink monochrome landscape tradition that was enthusiastically adopted by scholars and academy painters alike, spawning a host of competing schools. In Japan, however, the poetic colorist T'ang style was retained and its emotive potential was developed so far that eventually *Yamamoto-e* painting had little in common with either its Chinese contemporary style or its Tang sources".[20]

The word *Yamamoto-e*, in fact, means "painting of Japan," and its first understood examples can be seen in the murals at the Phoenix Hall at Byōdō-in. In these newly understood, purely domestically cultivated paintings, Buddhism (a common Chinese subject) was not necessarily an inspiration for a work. But in spite of the growing popularity of such secular works, the grip of Buddhism held firm, not only in works of art and landscape gardening, but in many other aspects of emergent individualistic Japanese culture, and was without rival for the next five hundred years.

20 Stanley-Barker, *Japanese Art*, 76.

Influences of Zen

It was the monk Eisai, who taught in Kyoto as early as 1202, who pioneered the distinctly Japanese interpretation of Chinese Buddhism that came to be known as Zen. The spread of this new interpretation of Buddhist thought grew steadily for the next 150 years as monasteries rapidly became established around the surrounding areas of Kyoto. The ruling families and the Shogun Takauji built Zen temples resulting in the doctrine of Zen coming into favor as a powerful aesthetic adaptation among the nobility. Within these temples grew Japanese arts and landscape gardens that were significantly distinct from past Chinese models. Indeed, as Seiroku Nouma notes, "in order to impress a rather broad and unsophisticated audience, Buddhist arts of the period had often stressed mechanical finesse and complexity in technique."[21] Zen Monks, particularly in the Kyoto area, sought to

"counter the debasement of the arts by eliminating ostentatious technique and the confusion of excellence with complexity; they sought to develop highly personal, direct forms of expression rooted in the most profound levels of Buddhist thought."[22]

These new forms of expression countered the established mediums through the use of *sumi-e* painting techniques and the gardening technique of *kare-sansui*. In *sumi-e* painting, dark monochrome was favored over the indulgence of color (prevalent in Chinese contemporary art), and brush strokes relayed simple construction that sought to express more intimate and revealing emotional reactions. In fact, the restrictions placed on the monks became a direct catalyst for them to recreate, in an abstract method, the understanding they had of the essence of reality. In using simplicity, the painters reflected the essence of Zen: that the eternal and incorruptible elements of reality are hidden beneath the complexities and chaos of the everyday world. They were able to transform "an ink painting tradition that had been transmitted to Japan from China centuries before."[23] The artistic premise centered on "evoking images of the evanescence of life, the vanity of human passions and a longing for an idealized afterlife."[24] In expressing a slim form, in monochrome, and with the use of large free areas of ink that often spread random blots, the beauty is revealed through space and the untouched areas of the paper. These first paintings, in their rough and lively power, were similar to calligraphic contemporary Zen scrolls or *bokuseki* (in-

21 Seiroku Nouma, *The Arts of Japan; Ancient and Medieval* (Tokyo: Kodansha International, 1966), 226.
22 Nouma, *The Arts of Japan*, 226.
23 Toshie Kihara, "The Search for Profound Delicacy: The Art of Kano Tan'yu," in *The Arts of Japan: An International Symposium*, ed. Miyeko Murase and Judith G. Smith (New York: Metropolitan Museum of Art, 2000), 83.
24 Rupert Faulkner, *Masterpieces of Japanese Prints* (Tokyo: Kodansha International, 1991), 11.

structional phrases or exhortations which were also sharp deviations from contemporary calligraphic practice). Simplicity and naturalness were key aspects to this art inspired by the Zen monks, characteristics that would permeate Japanese cultural and aesthetics predilections for centuries to come.

But notions regarding Zen and its philosophy towards the world also found an increasing influence in the design of a new garden prototype the *kare-sansui*. Literally meaning "withered mountain-water" the term first appeared in the mid eleventh century gardening treatise *Sakutei-ki* (Records of Garden making)[25] attributed to Tachibana no Toshitsuna (1028–1094). Initially the term referred to the inclusion of small rock groupings typical in Heian (794-1185) pond gardens, though by the beginning of the Muromachi (1336-1573) it had acquired a new meaning that strongly associated Zen ideals regarding emptiness and the void. As a dry landscape, yet more often than not signifying water, a *kare-sansui* is typically contained within a small space that is enclosed where a bed of white gravel is articulated with the placement of various rocks. Few trees or shrubs are present, if at all, with the use of moss, earth and the texture and shape of the feature rocks the main focal point of the garden.

Figure 2: The Muromachi period kare-sansui at Nanzen-ji, Kyoto

25 Jiro Takei and Marc P. Keane, *Sakuteiki Visions of the Japanese Garden: A Modern Translation of Japan's Gardening Classic* (Boston, Massachusetts: Tuttle Publishing, 2001).

They are then a radical departure from the Heian and Asuka paradise gardens of boating parties and the sumptuous suggestions of a mystical western land of Aimida Buddha. Reduced in scale and heightened in their austerity, they are inward looking, the white raked gravel bed suggesting a blank canvass while the solitary rocks drawing associations to mountaintops peaking through low clouds or islands rising from a serene sea. Indeed, the bed of gravel that literally grounds the composition of a *kare-sansui* is often raked into patterns that suggest waves or the movement of water. Rather than exploring the gardens through boating or walking though, the *kare-sansui* must be viewed from a seated position and often on a raised veranda from the *hōjō* or living quarters of the temple or monastery. The proliferation of dry gardens within Zen temples and monasteries of the time also suggests the shift away from the purpose of the Japanese garden from providing a space for aristocratic indulgence towards providing insights into the nature of Zen and its philosophy regarding awakening. As Thomas Hoover notes, the *kare-sansui*:

"Condensed the universe into a single span, and were primarily, if not wholly monochromatic. More importantly, they were intended exclusively for meditation. Whereas earlier landscape gardens had always striven for a quality of scenic beauty, these small temple gardens were meant to be a training ground for the spirit, a device wherein the contemplative mind might reach out and touch the essence of Zen."[26]

In this regard, what such gardens offered was an embodiment of the fundamental nature of Zen to present a "strong focus on the particular and the concrete aspects of ordinary experience as the site for the recognition of the persuasive presence of ultimate reality and of pervasive experience of enlightenment."[27] Similarly, there is a great concern with the illusory and empty nature of reality, which seems duly reflected in the design of the *kare-sansui*. Thus the austerity and simplicity of the dry landscape, devoid of opulence and outward grandeur places the onus on the viewer to interpret reality. If reality has become only an illusion, then to seek out the truth is to eliminate the superficial and look within in much the same way in which the visitor is confronted with the emptiness that seems to almost frame the rocks of a *kare-sansui*. This particular ideal is also evident in the closely related role that landscape painting and landscape gardening shared. In the developed works of Muromachi painter and garden maker Sesshū Tōyō (1420-1506) and the later works of seventeenth century painter Kanō Tan'yū (1602-1674), a very distinct form of ink painting arose that conveyed a Zen sensibility that could equally describe the approach to designing *kare-sansui*.[28]

26 Thomas Hoover, *Zen Culture* (London: Arkana, 1977), 102.
27 Joseph D. Parker, *Zen Buddhist Landscape Arts of Early Muromachi Japan (1336-1573)* (New York: State University of New York Press, 1999), 112.
28 Peter C. Swan, *A Concise History of Japanese Art* (Tokyo: Kodansha International, 1979), 44.

Figure 3: Mampuki-ji Teien, Masuda (attributed to Sesshu Tōyō)

During the Muromachi period, it was understood from the Chinese Masters that in ink painting, space was governed and appropriately constructed using a systematic division of the medium (paper/canvas). Objects that appear in the foreground of a painting are larger in size and darker in the quality of the ink than those represented in the background. In the hands of Sesshū and then later Tan'yū, the representation of space is particularly dramatic and highly unconventional; they clearly intended to create a sense of vastness within their landscapes. Indeed Sesshū is widely recognized as "responsible for developing stylistic innovations that became an important model for [later] sixteenth-century painters."[29]

A striking contrast to paintings of contemporary Chinese artists can be seen in the way that Tan'yū and Sesshū use large sections of open or empty space. These spaces may first be understood as clouds or fog: open structures and trees seem to poke out from the dense haze. Further consideration though shows that these large open spaces are the *truth* within the work. By this I mean that they represent only the illusion of reality, appearing to punctuate and help direct the real point of the work. They are, as Joseph Parker argues, a function of Buddhist thought and its conceptions about the relationship between humans and nature:

29 Michel-André Bossy, Thomas Brothers and John C. McEnroe, *Artists, Writers and Musicians* (Westport: Oryx Press, 2001), 167.

Figure 4: Sesshu Tōyō, Haboku-Sansui (1495)

"As a world of changing seasons and trickling streams, the natural world well represented to them the ever-transforming, unreliable, and even paradoxical or dream-like nature of sentient existence in the Buddhist view."[30]

Tan'yū's landscapes, much like the aesthetics of the *kare-sansui* "express two conflicting visions, namely the illusion of deep and wide spaces created by the rules of pictorial recession versus the unpainted areas, which permeate and penetrate separate parts of the painting and transform them into undefined surfaces."[31] These undefined surfaces must be realized by the viewer, and viewed individually; "the viewer is left to read the unpainted motifs and decide what the unpainted area represents."[32] Ink painting, like the dry landscape gardens of the Muromachi, sought to deliberately court "ambiguity, leaving empty spaces in their compositions for readers or spectators to fill in according to their intuit understanding of the ultimate meaning."[33] This is particularly evident in Sesshu's landscape painting *Haboku-Sansui* (Figure 4) in which all the elements of the painting are only suggestions for the viewer, not concrete representations. As Toshihiko Izutsu notes: "The whole landscape consists of indistinct forms, varying ink tones, vapors and the surrounding emptiness. In immense distances of the background, beyond veils of mist, craggy pillars of mountains looms against the sky, vague and obscure, like phantoms."[34]

Later Developments

By the beginning of the Edo period (1615-1867) in Japan, the rise of the Shogunate brought not only political stability, but also severe restrictions on overseas travel for citizens. But the traditional cultural and arts capital of Japan remained in Kyoto, where the Emperor was heavily subsidized by the Shogunate to construct numerous new gardens. As a consequence the Edo saw the rise of the *niwashi* or the professional garden maker who would eventually replace the role of the traditional designers and makers of gardens, Zen priests or monks.[35] But Zen Buddhist thought still remained a strong influence on the developments of arts and landscape design, though during the Edo, a new interest in Chinese Confucism arose. Though the Edo period is marked

30 Parker, *Zen Buddhist Landscape Arts*, 207.
31 Toshie Kihara, "The Search for Profound Delicacy: The Art of Kano Tan'yū," in *The Arts of Japan: An International Symposium*, ed. Miyeko Murase and Judith G. Smith (New York: Metropolitan Museum of Art, 2000), 63.
32 Kihara, "The Search for Profound Delicacy," 98.
33 William Theodore de Barry, *Sources of Japanese Tradition.* (New York: Columbia University Press, 2001), 365.
34 Toshihiko Izutsu, "Tokyo-Montreal: The Elimination of Color in Far-Eastern Art and Philosophy" in *The Realms of Color*, ed. Adolf Portmann and Rudolf Ritsema (Dallas: Spring Publications, 1974), 438.
35 Jiro Harada, *Gardens of Japan* (New York: Routledge, 2009), 6.

by the closing of the borders of the country to the outside world and a voluntary aversion to foreign influences, Zen monks at the end of the Muromachi era had already brought back from China texts of the philosopher Chu Hsi (1130-1200). In what Günter Nitschke describes as a particularly important influence on garden making, Hsi's "neo-Confucianism [was] chiefly the type offering a particular blend of Taoist mysticism and Buddhist metaphysics."[36]

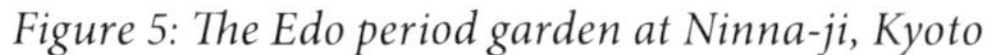

Figure 5: The Edo period garden at Ninna-ji, Kyoto

Many of the gardens of the Edo period thus moved away from the model of the small, self-contained *kare-sansui* dry landscapes that are viewed from a single seated position and attached to a temple, instead re-embracing the early prototype of the large pond garden. But the primary difference between the Edo stroll garden and earlier paradise or pleasure gardens is that Edo gardens tended to be much larger. This was due to the undeveloped nature of the areas such as Edo (modern-day Tokyo) where the *daimyo* (feudal lords subordinate to the Shoguns) had been assigned estates. But as David and Michiko Young note, these new large stroll gardens also combined traditional techniques in innovative ways to generate a new garden prototype:

"One technique (*shin-gyō-sō*) involves moving from a formal entrance through semiformal features to an informal interior. A second technique, *miegakure* (hide-

36 Günter Nitschke, *Japanese Gardens, right angle and natural form* (Köln: Tachen, 1999), 117.

and-reveal), uses hedges, structures and other objects to block long range views and provide an ever-changing vista as one turns a corner or crosses a bridge. A third technique, *shakkei* (borrowed scenery), involves incorporating a distant hill, mountain or pagoda into the garden view."[37]

Figure 6: The Edo period garden of Kenrokuen, Kanazawa

37 David Young and Michiko Young, *The Art of the Japanese Garden* (Tokyo: Tuttle Publishing, 2005), 124.

In addition to using the techniques of *shakkei* and *miegakure*, Edo stroll gardens were often also comprised of a series of *meisho*, or "famous views," that can be considered as fulfilling the role of a type of postcard of the era. As such, famous domestic or Chinese natural landscapes, like Mount Fuji or Mount Horai, or scenes from Taoist or Buddhist legends, or even landscapes illustrating famous poems were often represented.[38] As Lorraine Kuck notes, the Edo period was:

"The period of the 'Eight Views,' a concept originally derived from China [. . .] Copying this, eight famous poetic views had been listed of Lake Biwa, the large body of water near Kyoto, and it became the fashion to find these eight views, or others, in the garden."[39]

Thus unlike the Muromachi *kare-sansui*, the Edo stroll garden was designed to portray nature as it appeared. Rather than attempting to abstract rocks or topography into purely aesthetic or sculptural forms, Edo garden designers intended to be literal at all times. But this did not mean a complete abandonment of the dry landscape. Instead, gardens of this era sought to combine dry and wet, *Yin* and *Yang* in a way that harmonized and combined the techniques so as to provide realistic transitions and scenery. Through the experimentation of Kobori Enshu (1579-1647), the development of techniques for incorporating *o-karikomi*, or aesthetically trimmed shrubs and small bushes, reached a pinnacle such that this new form of topiary played a central role in complementing particular scenes within the garden. Exquisitely trimmed shrubs were used to suggest clouds or mountains or as complementary forms to signifying Buddhist rock trios, or the heavily referenced crane or turtle islands. But other *kare-sansui* style gardening techniques also became incorporated in the larger Edo gardens. In contrast though to the Heian and Asuka pond and island gardens, the Edo garden focused heavily on the use of rocks, particularly in the design of pond shorelines, beaches, ravines and islands. Though these rocks settings used significantly larger rocks (often representing status symbols of their owners), their use within group compositions was often simpler than in earlier times. This was to better aid a view of the garden from higher up than that experienced by visitors traversing through the garden by foot. As Nitschke notes "perhaps it was the view from the top of the new fortified palaces that inspired the Japanese to approach their gardens from a bird's-eyes perspective."[40]

Perhaps the three key aesthetic ideals of the Japanese garden that began in the Momoyama era (1568-1600) and fed through to the late Edo and later Meiji period (1868-1912), that of *wabi*, *suki*, and *sakui*, can best describe the subtle change in

38 Phillip Pregil and Nancy Volkman, "Landscapes of the Rising Sun: Design and Planning in Japan," in *Landscapes in History: Design and Planning in the Eastern and Western Traditions* (New York: John Wiley & Sons, 1999), 363.
39 Lorraine Kuck, *The World of the Japanese Garden* (New York: Weatherhill, 1968), 243.
40 Nitschke, *Japanese Gardens*, 118.

attitudes towards garden design and the function of the garden. *Wabi*, which Miyeko Murase describes as a "sober refinement and calm,"[41] is most associated with the tea ceremony (*chanoyu*) and the small gardens, or *roji*, that were situated next to the tea house (*chashitzu*). The first great tea master Sen no Rikyū (1522-1591) introduced the strict guidelines for the design of the *roji* such that it should emanate complete restraint. The extremely plain and simple *chanoyu* was to mimic a hermit's hidden mountain hut—a small wooden structure with a thatched roof. The garden followed the notion of *wabi* in that there is an ordinary quality to its simplicity though they were constantly watered and damp to bring vibrancy to the greenery of the moss and shrubs. Flowering plants and highly contrastive colors were avoided so as to not distract visitors.

Figure 7: The Taishō garden Kyu Furukawa Teien, Tokyo

In concert with the aesthetic ideal of *wabi* which Haga Kōshirō describes as finding the deeper beauty "in the blemished [rather] than in the unblemished,"[42] is *suki* which represents connoisseurship and *sakui* which refers to personal creativity.[43] Thus the tea garden became a model of contemporary attitudes towards presenting the ordi-

41 Miyeko Murase, *L'Art du Japon* (Paris: La Pochothéque, 1996).
42 Haga Kōshirō, "The *Wabi* Aesthetic through the Ages," trans. Martin Collcuttin, in *Tea in Japan: Essays on the History of Chanoyu*, eds. Paul Varley and Kumakura Isao (Honolulu: University of Hawaii Press, 1989), 198.
43 Marc Peter Keane, *Japanese garden design* (Tokyo: Tuttle Co., Inc., 1996), 76.

nary and even the artless as a means to withdraw from the extravagance of the world and to present a new space of austerity while at the same time allowing for small touches of individual taste from the garden designer (who was also a tea master). But by using elements common to previous garden prototypes such as *sawatari* (stepping stones) in combination with rustic looking stone lanterns, the atmosphere of the *roji* is presented as a means to aesthetically engage the visitor who was inevitably on their way to the tea ceremony. Thus the function of the *roji* changed significantly from a self-referential garden of previous eras and prototypes to a transitional and functional space whose main goal was in aiding in the construction of a suitable atmospherics of subtlety and of *wabi* that would accompany the simplicity and austerity of *chanoyu*.

But by the beginning of the Taishō era (1912-1926) a large number of historic gardens of the Edo and Momoyama eras had run into such disrepair through neglect and abandonment that many were designated for restoration so as to convert them into public parks. In an effort to continue the opening up and modernizing of Japan after the Meiji restoration of 1854, government officials employed foreign trained architects such as Josiah Conder and Wilhelm Böckmann (as *oyatoi gaikokujin* or foreign advisors) with the task of the restoration of these sites and the widespread incorporation of Western style architectural building projects within Japan. This sudden heightened influence though also caused what Marc Treib and Ron Herman suggests as "strange concoctions of Western forms and Japanese building techniques normally used for castles and storehouses to spring up in the foreigners' residential districts."[44] But new approaches to garden design were also initiated. The interface between Western gardening techniques and traditional Japanese garden aesthetics produced a number of mixed approaches to garden management and restoration. But such a synthesis between Western and Japanese ideals regarding nature and the garden also produced in the Taishō era a rise in the number of private gardens designed and built, marking a throwback to early Heian models of gardens located within a residence rather than attached to a temple. But instead of also drawing on the Heian concept of the role of the paradise garden, the Taishō garden followed on from the Edo and Meiji notion of presenting nature as it appeared, though in a much stricter fashion. As Nitschke notes:

"Gardens were now expected to be truthful copies of nature in its 'real' form. They were no longer 'nature as art,' nature designed and molded by human hands, but simply a part of nature made by nature. The selective, reductive, abstractive hand of the designer was to remain hidden so that the garden might appear a perfect icon of nature."[45]

The influence of the West regarding naturalism was a trend not only in introduced into Japanese gardening but similarly found traction within contemporary art practi-

44 Marc Treib, *A Guide to the Gardens of Kyoto* (New York: Kodansha, 2003), 33.
45 Nitschke, *Japanese Gardens*, 210.

ces such as in the work of printmaker Kobayashi Kiyochika (1847-1915) and the painters Mori Ogai (1862-1922) and Natsume Soseki (1867-1916) who had both studied in the West. But with the loosening of the traditional ties of landscape gardening in Japan to the historical codification of all elements of a garden—from its topographic manipulations, rock formations and architectonic features, etc.—the influence of Western norms of gardening found numerous additions to traditional garden design including the use of flat grassy lawns and a greater variety of exotic plant and tree species. This loosening though was also accompanied by a sharp increase in the publishing of garden manuals, many of them written by layman or amateurs, such that there was a sharp decrease since at least the Edo period of relevant scholarship and research in and around Japanese garden aesthetics and the techniques of garden design and making.

* * *

What I have sought to present in this first chapter is an introduction to some of the core elements and historical events and influences that have shaped the progression of the Japanese garden from its early Chinese adaptations of the Heian period to its later independence and further evolution into and beyond the Meiji era. As I previously mentioned, the Japanese garden has not only been a model for contemporary designers such as landscape architect Mirei Shigemori's numerous temple gardens or the Koshino House by architect Tadao Ando, but my own creative works of which I will be exploring in detail throughout this book. This introduction has specifically focused on garden aesthetics and spatial characteristics common in gardens up to the Taishō era as a means to provide a background for later investigations in this book that deal with specific gardens from those periods, in particular the Muromachi dry landscapes of Sesshutei and Ryōan-ji, Edo era Koishikawa Korakuen and Taishō garden Kyu Furukawa Teien. But as a means to access this body of creative work, I will in the next chapter examine the musical composition *Ryoanji* by American composer, thinker and iconoclast John Cage (1912-1992). I position Cage's work in translating the Muromachi-era *kare-sansui* garden at the temple Ryōan-ji in Kyoto as an important precedent for my own investigations into the Japanese garden as a *spatial model.* By examining not only the method by which Cage produces his musical score, but also the aesthetics of the dry garden and indeed Cage's understanding and reading of the qualities of the space, enables a clearer picture to emerge of what techniques and approaches are required to usurp the spatial proclivities and multi-sensory encounters afforded visitors to a Japanese garden. To examine Cage's work as a paradigm then gives a clear central theme to the larger aim of this book which is to explore, document, analyze, and via artistic and creative endeavors, to *think through* the Japanese garden in order to understand both its physical and metaphysical dimensions.

John Cage's Ryoanji

A Case Study for Spatial Thinking

> The purpose of art is to imitate nature in all her manner of operations.
> JOHN CAGE, *SILENCE*, 100.

CAGE'S AESTHETICS

The influence of Zen Buddhist thought on the musical aesthetics of American composer John Cage (1912-1992) remains a most recognized element in a career for which lure of indeterminacy provided the most suitable means for constructing a musical model of the natural world. Throughout Cage's career as composer, poet, artist and thinker, his championing of experimental approaches, sustained experimentation and self-documentation of the removal of composer intention is a legacy of his first introduction and later adaptation of the Chinese oracle I-Ching in 1951 through colleague Christian Wolff. This critical juncture heralded a paradigm shift in Cage's compositional approach, expanding its domain to produce numerous other branching aleatoric methodologies that developed over the latter course of his career. These techniques were diverse and ranged from the fashioning of thematic materials by observing imperfections in blank manuscript paper, transcribing star charts printed on transparencies, using computer software for randomization strategies or generating data-sets through coin tossing to guide musical form. But Cage's embrace of such processes were principally as a means to sustain his notion that music should mirror "nature in all her manner of operations,"[1] thus placing Cage's musical aesthetics in defiance of those contemporary expectations on the relationship between inner compositional intention and the performer's sensuous musical interpretation of that intention—what philosopher Theodor Adorno saw as the technê of musical produc-

1 James Pritchett, *The music of John Cage* (Cambridge: Cambridge University Press, 1993), 91. See also Cage discussing Coomaraswarmy in John Cage, *Silence* (Middletown CT: Wesleyan University Press, 1961), 194.

tion and musical re-production.[2] That Cage was introduced to Japanese thought a year after his acquisition of the I-Ching through the Zen scholar Daisetz Suzuki's tenure at Columbia University from 1952-1957 is perhaps the composer's most readily documented anecdote—for which its legacy is widely utilized in numerous works of music, poetry, printmaking and performance art.[3] As a self-confessed experimentalist then, Cage routinely challenged prevailing post-war musical conventions by radicalizing audience expectations. He primarily achieved this through questioning the traditional methods by which music should be composed and delivered, while at the same time identifying the utilitarian nature of his avant-garde explorations as potential models for new social paradigms. Cage's assertion that "everything I do is available for use in the society"[4] became an important part in his efforts to drive nineteenth-century musical expectations beyond their own apparent self-enforced confines and reliance on goal-oriented outcomes. By the middle of the 1950s, Cage's embracing of a new Zen-derived model for conceptualizing sound and sound making produced a new radical aesthetics that sought to systematically reconnect common auditory experiences of the everyday to those rarefied experiences of the concert hall. This was inevitably achieved through what Noël Carroll describes as a framework of ex hypothesi, where audience coercion attended to the sounds of a work. Carroll argues that:

"Cage's goal is to design a context in which *ex hypothesi*, the audience is, in a manner of speaking encouraged, if not coerced or reduced, to attending to the qualities of the sounds themselves for the simple reason that the events have been constructed in such a way that there is nothing else to which one could attend."[5]

But for Cage, innovative compositional directives also became devices for the recontextualization of sound, which when treated as a material, delivered striking new textures and sculptural possibilities. Similarly, his use of found objects, unconventional instruments, novel thematic materials and everyday behavioral patterns and gestures provided a means for which the perceived distinction between the performing space of music and the performing space of the everyday becomes indistinguishable. However, Cage's radicalization of inherited nineteenth century musical aesthetics was also delivered through the reach of Zen beyond its own discourses. His visit to Japan in the early 1960s, and in particular to the famous dry rock gar-

2 Theodor Adorno, *Sound Figures*, trans. R. Livingston (Stanford, CA: Stanford University Press, 1999), 197.

3 See in particular, Cage reading numerous short stories and anecdotes inevitably concerning Zen aesthetics (with David Tudor performing piano and live electronics) in *Indeterminacy: New Aspect of Form in Instrumental and Electronic Music: Ninety Stories by John Cage, with Music*, Smithsonian Folkways FT 3704 (2 LPs).

4 Joan Retallack, *Conversations in Retrospect* (Hannover, NH: University of New England Press, 1996), xxvii.

5 Noël Carrol, "Cage and Philosophy," *The Journal of Aesthetics and Art Criticism* 52/1 (1994): 94.

den of the temple Ryōan-ji in Kyoto, delivered a vital impetus for a methodological apparatus that sought to compose-out the garden's spatial structure. Though the definitive musical endpoint of his visit to the garden would only come some 20 years later in the form of the oboe and obbligato percussion work, *Ryoanji*, the tenets of the garden's spatial relationships and Cage's reading of them provide a valuable model for interrogating the spatio-temporal dimensions of his musical aesthetics.

Cage and the Garden at Ryōan-ji

The *kare-sansui* at the Rinzai Buddhist temple of Ryōan-ji in Kyoto is universally recognized for its simplicity of form and austerity of materials (Figure 8). As Cage would have experienced in his first visit there in 1961, the garden is situated at the southern edge of the *hōjō*, or the abbot's quarters, for which a low veranda borders one of two leading edges of a large 15m x 6m rectangular area. A low earthen wall encloses the space that is primarily comprised of meticulously raked gravel. Within the gravel are five groups of rocks embedded within islands of moss, fifteen in all, though for the viewer at any point along the low veranda, only fourteen of these can be seen at any one time. A product of the Muromachi, the garden's intangible origins (it was constructed by anonymous *kawaramono* or garden workers in the late fifteenth century) only seems to compliment the continuous mystery of its meaning.

Figure 8: Kare-sansui garden at Ryōan-ji, Kyoto

Western readings of the space as seemingly quasi-modernist in its sculptural qualities[6] only seem to alienate even further the obvious divide that the garden's original designer presents contemporary viewers, whose current separation from Muromachi aesthetics becomes particularly accentuated at the site. It is perhaps the legacy of this uniqueness that drove Cage to re-compose the garden as a musical score. Using a collection of his own fifteen stones, he traced around their edges as a way to generate a melodic slide between chance-derived pitches (played as glissandi):

"Paper was prepared that had two rectangular systems [musical staves]. Using two such sheets I made a "garden" of sounds, tracing parts of the perimeters of the [. . .] stones [. . .] I was writing a music of *glissandi*. [. . .] For the accompaniment I turned my attention to the raked sand. I made a percussion part having a single complex of unspecified sounds played in unison, five icti chance-distributed in meters of twelve, thirteen, fourteen or fifteen. I didn't want the mind to be able to analyze rhythmic patterns."[7]

Cage's translation of the *kare-sansui* into a musical analog contained within two rectangular sheets was a direct mirror of the spatial boundaries of the original garden. His re-imaging of the garden as a musical space housed within a rectangular manuscript stave directly identifies the potential for proportional relationships as an aid to translation (Figure 9). Essentially, Cage uses the spatial proclivities of the garden's geometry as a metric and a valuable inter-media transfer tool.

Figure 9: Ryoanji (1983), John Cage, page 8 (Copyright © 1985 by Henmar Press, Inc.)

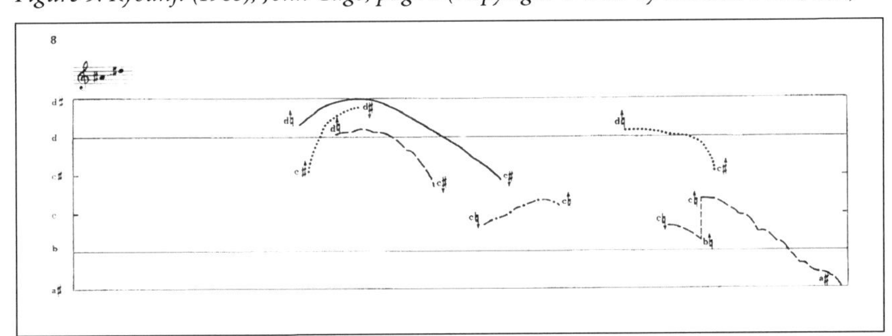

6 Thomas Hoover, *Zen Culture* (London: Arkana, 1977), 111.
7 John Cage, "*Ryoanji*: Solos for Oboe, Flute, Contrabass, Voice, Trombone With Percussion or Orchestral Obbligato (1983–85)," *A Journal of Performance and Art* 31/3 (2009): 57–64.

The method situates the articulation of the *kare-sansui* within its own container as an analogous model to the articulation of notation within the container of a musical score. Thus the suggestion of the garden as a type of rhythmic score, in which points on a plane facilitate a proportional notation, would have mirrored numerous extent graphic works of Cage such as *Music for Carillion No. I* (1952) and *Variations I* (1958). Indeed, David Slawson argues that the geometry and placement of the rock groupings at Ryōan-ji works as a counter to the perceived stasis of the garden, an observation Cage may well of understood when he initiated his own musical translation. Slawson elaborates:

"The spacing between the rocks can nevertheless produce a visual sensation as intriguing and intense as any arising from the character of the rocks themselves. The effect can be further heightened when the rocks also establish a complementary pattern of movement, or thrust."[8]

But in terms of the history of Japanese gardening aesthetics, this containment and demarcation of the *kare-sansui* into a functional space viewed from a single seated position on a temple veranda was a significant innovation that broke with contemporary conventions. Within the Muromachi *kare-sansui*, the landscape becomes abstracted rather than represented, and though operable space is physically reduced, an illusion of a vast almost cosmic scale of events emerges.[9] Are the fifteen rocks at Ryōan-ji islands in a sea (the raked gravel suggesting waves) or perhaps mountain peaks penetrating low clouds? The removal of a referential landscape as a context for the *kare-sansui* means that the focus for the visitor turns immediately inwards: towards the rocks and the empty space between them.[10] By default, Cage's musical translation of the *kare-sansui* similarly focuses our aural attention onto the constituent elements of the garden—the rock groupings and the raked gravel—though now not as a vehicle for analyzing their musical qualities as structurally polyphonic or thematically linked. The 'empty' space between the rocks is transformed into a type of musical stasis that is neither background nor foreground: the obbligato percussion part is regular and repetitive though designed to be perceived as a-periodic—a seemingly unchanging ground. Cage requires the materiality of the percussive instruments to be:

"At least two only slightly resonant instruments of different material (wood and metal, not metal and metal) played in unison [. . .] These sounds are the 'raked sand' of the garden. They should be played quietly but not as background."[11]

8 David A. Slawson, *Secret Teachings in the Art of Japanese Gardens* (New York: Kodansha, 1987), 95.
9 Mark Holborn, *The Ocean in the Sand* (London: Gordon Fraser, 1978), 58.
10 Lorraine Kuck, *The World of the Japanese Garden* (New York: Weatherhill, 1968), 164.
11 John Cage, *Ryoanji* for oboe and obbligato percussion (New York: Henmar Press, 1983), 1.

By using chance operations transcribed from the I-Ching, both the rhythmic structure of the obbligato percussion part and the melodic density of the oboe part were generated. Cage's embrace of aleatoric methods within the development of his compositional systems served primarily to remove pre-conceived musical expectations. The I-Ching or *Yì Jīng*, though originally a text for divination was used by Cage to generate random number sequences that could be utilized in constructing compositional form, pitch information or durational information for a work. By tossing coins, six either broken or unbroken lines are used to construct a vertical hexagram of which there are sixty-four possible combinations (2^6). These sixty-four hexagrams are numbered and named within the I-Ching and were originally used to answer questions put to the oracle in advance.[12] Cage's re-application of the oracle occurred through the asking of questions in regards to specificities of thematic materials (for example, how many sounds to occur in a work), instrumentation (which instruments are to be used) or general structural traits (what is the melodic range, dynamics or timbres). By using chance methods, outcomes at the time of composition were unknown and musical goals regarding the shape or texture of the thematic materials were abandoned in favor of embracing all possibilities.

Figure 10: Ryōan-ji from Miyako rinsen meishō zue (1799) Ritoh Akisato

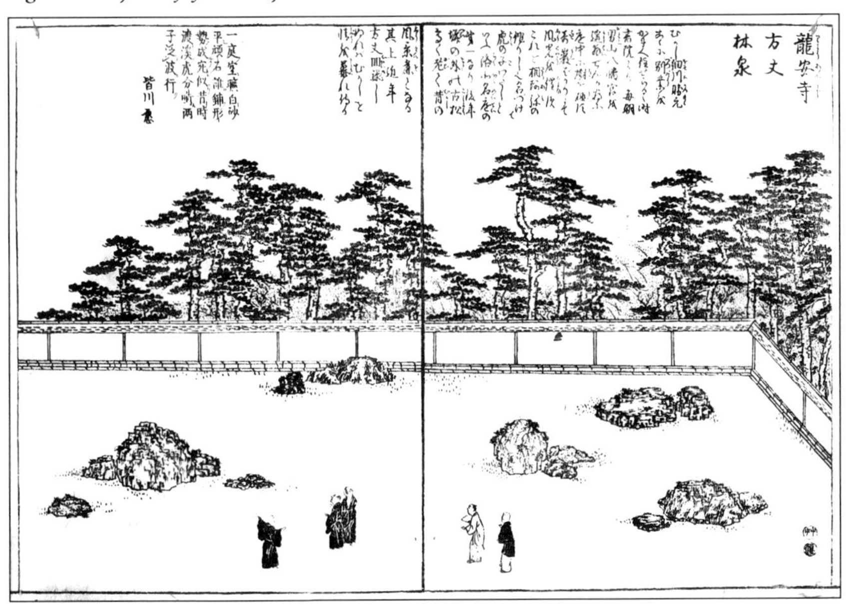

12 Hellmut Wilhelm and Richard Wilhelm, *Understanding the I-Ching: The Wilhelm lectures on the Book of Changes* (Princeton: Princeton University Press, 1995), 122.

In the case of *Ryoanji*, aleatoric principles also provided a means in which the musical form of the work becomes open. Conceived initially as an iteration of the two rectangular manuscript containers representing the garden as its smallest indivisible fundamental, *Ryoanji* may be performed as any number of the seventeen pages of the score played in any order. That the resultant form might be hidden even from the composer creates a model of the unknown in which the act of composition becomes a catalyst for generating new listening modes. But Cage further accentuates the novel listening space of *Ryoanji* by designating the rectangular containers of the manuscript pages as a garden of sounds that are "to be played smoothly and as much as possible like sound events in nature rather than sound in music."[13] That the melodic glissandi represent the stones themselves and are planted within their containers as agents that simply coexists with the sounds of the percussion also casts the work as one that seeks to de-border traditional musical hierarchies. The stipulation that the score represents "a 'still' photograph of mobile circumstances"[14] suggests the importance Cage placed on *Ryoanji* as unfixed in its spatio-temporal dimensions. That a vast number of temporalities be operable between performance versions of *Ryoanji* seems only to mirror the nature of the *kare-sansui* to question the appropriate applicable scale when viewing the garden. Perhaps playing with such an ambiguity led the Edo garden scholar and artist Ritoh Akisato to portray the *kare-sansui* at Ryōan-ji in such stunning detail within his six-volume monograph *Miyako rinsen meishō zue* (1799). Depicting what Gert van Tonder identifies as those "canonical views of idealized scenery [. . .] where the garden is intended to be seen from a preferred viewing location,"[15] Akisato includes four priests and a samurai who seem to be discussing the rock settings within the garden itself (Figure 10). In a curiously stylized manner the figures have been miniaturized and painted at a comparable scale to the feature rocks.

But if questions of scale seem to remain indeterminate, it is perhaps the connection of the garden to the veranda of *hōjō* that remains the nexus to the figure of the rock groupings. Indeed, Michael Lyons contends that the underlying spatial structure of the rock groupings and their relative articulation within the rectangular container is a result of a highly considered design approach. By using Medial Axis Transformation (MAT) a natural tree-like branching structure or tessellation inherent in the empty spaces between the rock groupings at the *kare-sansui* at Ryōan-ji is revealed in which "medial axes do not converge towards the viewing area upon deletion or addition of the clusters, [. . .] thus the structure defined by the shape of the ground in the original layout of Ryōan-ji is non-accidental."[16] The resultant composition then manages to locate the viewing area on the veranda as the prime point of visual convergence of the site—an effect that masterly connects viewer to garden and temple architecture.

13 Cage, *Ryoanji*, 1.

14 Cage, *Ryoanji*, 1.

15 Gert van Tonder, "Recovery of visual structure in illustrated Japanese gardens," *Pattern Recognition Letters* 28 (2007): 728-739.

16 Gert J. Van Tonder and Michael J. Lyons, "Visual Perception in Japanese rock garden design," *Axiomathes* 15 (2005): 367.

The overwhelming impact of the empty space extant between the feature rock groupings also seems though to readily communicate the Japanese concept of *yohaku no bi*, the technique which artists such as Sesshu and Tan'yū greatly exploited in their contemporary ink paintings. Literally referring to the beauty of empty space, Marc Peter Keane notes "this terse aesthetic can be perceived in the sparse ink painting of the [Japanese] medieval period with their expanses of unpainted paper as well as in the large empty spaces found in the gardens of the time."[17] At Ryōan-ji, it is the scale and intervals between the rocks that necessarily generate the larger shape of the whole composition. For Günter Nitschke it is this emptiness in the rock arrangement that provides the most profound impact. Observing the effect of the garden as an aid for meditation, Nitschke identifies the non-referential character of the *kare-sansui*:

"It needs the sophisticated interplay of form with its non-form, of object with its space. It is here, perhaps, that we find the ultimate purpose of garden art—to provide the necessary forum for such insight. The garden of Ryōan-ji symbolizes neither a natural nor mythological landscape. Indeed, it symbolizes nothing, in the sense that it symbolizes *not*. I see in it an abstract composition of "natural" objects in space, which is intended to induce meditation. It belongs to the art of the void."[18]

Figure 11: Kare-sansui garden at Ryōan-ji, Kyoto

17 Keane, *Japanese garden design*, 57.
18 Nitschke, *Japanese Gardens*, 92.

As a particularly important guiding principle of Muromachi era aesthetics, *yohaku no bi* is bound to the technique used in both gardening and painting in which emptiness, or white space, becomes a desirable property in, and of, itself. The reading of empty space as a balancing force that evokes the apogee of beauty may also open a scene to the perception of a cosmic scale. A characteristic that Thomas Hoover suggests is uniquely inherent in Ryōan-ji, where the garden "evokes a sense of infinity in a confined space [. . .] a living lesson in the Zen concept of nothingness and non-attachment."[19] But writing in the first edition of his publication *Silence* in 1961, Cage also observes the necessity of reading such emptiness within Ryōan-ji as relatively defined through its non-emptiness:

"For it is the space and emptiness that is finally urgently necessary at this point in history (not the sound that happens in it—or their relationships) (not the stones—thinking of a Japanese stone garden—or their relationships but the emptiness of the sand which needs the stones anywhere in the space in order to be empty)."[20]

This reading of the relationship of coexistence between objects and space within the *kare-sansui* exemplifies in Buddhist terms what Graham Parkes describes as "the principle of interdependent arising (Jpn. *engi*, Skt. *pratītya samutpāda*), the idea that each particular is what it is only in relation to everything else."[21] But Cage's treatment of the melodic foreground and percussive background sounds of *Ryoanji* also reveals an affinity for the manner in which philosopher and sociologist Henri Lefebvre challenges normative representations of absolute space. Cage's contention that "there is no such thing as an empty space or an empty time"[22] implicates the composer's compositional process as one that addresses the inert quality of music as spatio-temporal phenomena. That music should be channeled through a listening mode that de-emphasizes what Adorno considered the functional, that is, the manner in which premeditated form, structure and sonnified thematic materials are meaningfully organized in a work, primarily serves to emphasize Cage's new musical paradigm. Adorno's chief criticism then of Cage's compositional approach relates directly to the manner in which music had been traditionally constructed, that is, through the temporal and thematic unfolding of sonnified dialectic structures.[23] As Adorno states, "relations in their turn, as the incarnation of the subjective dimension, cannot be regarded as the primal material of music: there are no notes without relations, no relations without

19 Hoover, *Zen Culture*, 110.

20 John Cage, *Silence: Lectures and Writings* (Middletown: Wesleyan University Press, 1973), 70.

21 Graham Parkes, "Further Reflections on the Rock Garden of Ryōanji: From *Yūgen* to *Kire-tsuzuki*" in *The Aesthetic Turn: Reading Eliot Deutsch on comparative philosophy*, ed. Roger T. Ames (Chicago: Open Court Publishing Company, 1999), 19.

22 Cage, *Silence*, 8.

23 Brandon W. Joseph, "John Cage and the Architecture of Silence," *October* 81 (1997): 80-104.

notes. Deception is the primary phenomenon. The hypostatizing of relations would be the victim of exactly the same myth of origins as the reduction to the naked note, but in reverse."[24] But for Cage, much like Lefebvre, there must be *indifference* between formal content (or musical sounds) and material container (or traditional thematic relationships):

"We know that space is not a pre-existing void, endowed with formal properties alone. To criticize and reject absolute space is simply to refuse a particular *representation*, that of a container waiting to be filled by a content – i.e. matter, or bodies. According to this picture of things, (formal) content and (material) container are *indifferent* to each other and so offer no graspable difference. Any thing may go in any "set" of places in the container. Any part of the container can receive anything. This indifference becomes separation, in that contents and container do not impinge upon one another in any way. An empty container accepts any collection of separable things, separate items."[25]

Cage's use and understanding of silence in music as not an absolute but a relative term pertinently embodies Lefebvre's notion of the principles of space as a matrix of content and container. The composer's oft quoted anecdote of his 1951 experience of 'silence' at the Harvard anechoic chamber encourages not only Cage's imbuing of non-intention as a compositional inevitability, but similarly the dissolving of those peculiarities between locative thematic materials existing as either *inside* or *outside* of music-space:

"There is always something to see, something to hear. In fact, try as we may to make silence, we cannot. For certain engineering purposes, it is desirable to have as silent as room as possible. Such a room is called an anechoic chamber, its six walls made of special material, a room without echoes. I entered one at Harvard University several years ago and heard two sounds, one high, one low. When I described them to the engineer in charge, he informed me that the high one was my nervous system in operation, the low one my blood in circulation. Until I die there will be sounds. And they will continue following my death. One need not fear about the future of music."[26]

When Cage defines silence as simply those sounds that occur unintentionally, the borders and demarcation of music-space as a controlled interior condition are softened. That which traditionally can be defined as noise, background sounds or interruptions

24 Theodor W. Adorno, "Vers une musique informelle" *Quasi una fantasia: Essays on Modern Music*, trans. Rodney Livingstone (London: Verso, 1998), 301.
25 Henri Lefebvre, *The Production of Space*, trans. Donald Nichols-Smith. (Oxford: Blackwell Publishing, 1974), 170.
26 Cage, *Silence*, 8.

to the performance of music now become part of the sound-space—a necessity of the indifference when reading sounds as both thematically intentional (interior) and non-intentional (exterior). That this paradigm also resembles the relationship between foreground and background in *Ryoanji* arises from manner is which Cage's process of translation seems to dismantle the work's reliance on hearing hierarchies in musical sound-space—each element coexists at any point within the musical container and their relational patterns neither produce nor impinge on any predefined musical expectation of a fixed hierarchy. Removing what Adorno saw then as the normative associations of musical meaning—that is, as a function of the structural patterns arising from integrated thematic elements—allows Cage to freely conceptualize the work as an auditory garden in which the juxtaposed sounds operate as autonomous agents within the container.

Additionally, Cage's reassessment towards reading the emptiness of the *kare-sansui* at Ryōan-ji as void also suggests a codification of his stance regarding the reconciliation between intention and non-intention:

"Just as I came to see that there was no such thing as silence, and so wrote the silent piece [4'33"], I was now coming to the realization that there was no such thing as nonactivity. In other words the sand in which the stones on a Japanese garden lie is also something."[27]

Such an observation, essentially on the qualities of *yohaku no bi* within Ryōan-ji, parallels what David Slawson notes as the "shift in emphasis from a feature-oriented to a quality-oriented aesthetic in garden making"[28] between the Heian and Muromachi eras. The emptiness at the *kare-sansui* at Ryōan-ji is a non-representational element, yet an important structural key that underlies the entire composition. While the garden is non-representational in terms of directly imitating a foreign or ideal landscape, its success lies in the space between the feature rock groups as a device that is highly suggestive and expressive, and one that inevitably draws the viewer into the composition as cohabiter—a nothingness that becomes *something*. The emptiness of the acoustic space at the Harvard anechoic chamber is perhaps the link then that connects Cage's views on non-action regarding performance practice to *yohaku no bi*—within the chamber, and even in complete physical stillness, sound perpetuates unintentionally.

27 Christopher Shultis, "Silencing the Sounded Self: John Cage and the Intentionality of Nonintention," *The Musical Quarterly* 79/2 (1995): 322.
28 Slawson, *Secret Teachings*, 72.

Further Thoughts

What is also evident though in Cage's understanding between action and non-action is his stance towards the role between audience and composer. Some of Cage's later work sought to directly engage and draw in the audience by inviting them to identify what constitutes the thematic materials of a work. In a piece such as *Demonstration of the Sounds of the Environment* (1972), the subtle connection between site, silence, audience and new listening modes becomes seamlessly integrated as a social phenomenon. Describing the original directives of the work for three hundred listeners, Cage recalls that:

"Through *I Ching* change operations we subjected a map of the university campus to those operations and made an itinerary for the entire audience which would take about forty-five minutes to an hour. And then all of us, as quietly as possible, and listening as attentively as possible, moved through the University community".[29]

Representing a structural précis of Cage's necessity for considering compositions in sound-space as dynamic, social entities, the work generalizes those engaged site interrogations in other works such as *Ryoanji*. Acting as a generator of a spatio-temporal ontology, *Demonstration of the Sounds of the Environment* defines the relationship between spaces and sound for which those other manifestations each become taxonomic class elements. Cage generalizes the concept of site in the work as a means to reveal the importance of sound as a marker of identity, articulator of space or marker of place. But as a social event, *Demonstration of the Sounds of the Environment* also highlights the composer's desire for listening within a site to be an interpenetrating consequence of embracing any sound source as a musical event.

Perhaps this leads directly back to notion that for Cage, if the sounds of the environment are to be understood as musical, than this approach is only effective if those sounds be observed *in situ* and operating as natural functions of the landscape. This notion seems congruent with his assertion that "music as I conceive it is ecological. You could go further and say that it IS ecology".[30] Certainly there is here an implication that functional and coherent sonic objects will always belie their designer's intent. By this I mean that unintentional sounds are *revealed* by the site's context, performers and audience as sonnifications of nature "in all her manner of operations." This revealing of nature as a series of structural relationships and patterns that become sonnified and recontextualized is a feature that Nitschke identifies as correspondingly indigenous to Japanese garden design:

29 Hans G. Helms, *John Cage talking on music and politics*, Watershed Tapes C-402A-cassette tape. See also Susan McClary, *Feminine Endings: Music, Gender, and Sexuality* (Minneapolis: University of Minnesota Press, 1991), 111.
30 John Cage, *For the Birds* (Hanover, NH: Wesleyan UP, 1995), 229.

"The Heian garden imitates the outer forms of nature within a selective landscape of natural features. It seems to me that the Muromachi garden takes a step further: it seeks to imitate the inner forms of nature and thereby fathom the secret laws of its proportions and rhythms, energy and movement. Its means an abstract composition of naturally occurring materials [*sic*]. Nor is there anything 'unnatural' about such compositions; after all, their rocks came directly from nature, where they would have remained unseen but for the detective eye of the designer".[31]

By using the lens of imitating nature, Cage's music seems equally concerned with both the direct usurpation of outer natural forms; say, as the surrounding sound world in *Demonstration of the Sounds of the Environment*, or those inner spatial forms; that is, as geometric and topographical patterns in *Ryoanji*. In presenting such an aesthetic framework, Cage's desire for music to be removed of subjective evaluations resonates with what Allen Carlson considers as the ability of the Japanese garden to reconcile the dialectical relationship between the natural and the artificial:

"Japanese gardens [. . .] [follow] the lead of nature rather than art. However, they do so not by means of making [the] artificial unobtrusive, but rather by making the natural appear in such a way that the tendency to judge is again averted. In short, Japanese gardens solve the problem of the role of critical judgment by rising above judgment in a way similar to the way nature itself does".[32]

Cage's stance on value judgments within musical composition similarly centers on his desire to use the experience of sound events (either within everyday life or rarefied artistic contexts) as a means to move beyond critical judgments. He argued that value judgments are "destructive to our proper business which is curiosity and awareness [. . .] We waste time by focusing upon these questions of value and criticism and so forth and by making negative statements. We must exercise our time positively."[33] But Carlson's further contention that the Japanese garden is neither pure art nor pure nature, achieving an a-harmonious nexus between the two also resonates with the wider *modus operandi* of Cage's compositional methods, while also pointing towards the notion of *mono no aware* as such a central theme to the greater aesthetic goals of Japanese garden design. That Cage preferred though to construct his sound-space via chance methods and thus alleviate knowledge of its final design form sits in obvious opposition to the extremely controlled interventions extent within Japanese garden design. Nevertheless, when music becomes conceptualized as ecological through the artifices of composer, performer or audience-led facilitations, it indeed approaches those same dialectics that are extant within the Japanese garden—that is, between the

31 Nitschke, *Japanese Gardens*, 106.
32 Allen Carlson, *Aesthetics and the Environment, the appreciation of nature art and architecture* (London: Routledge, 2000), 169.
33 Richard Kostelanetz, *The Theatre of Mixed Means* (London: Pitman, 1970), 57.

intervening hand of the designer and the natural form and features of the landscape. This relationship between intention and non-intention, sound and silence, music and nature becomes the structural base for which Cage's subtle interrogations reveal hidden patterns, tensions and unexpected encounters. His works then can be read as neither wholly functional as meaningful in the context of traditional music-space aesthetics, nor as strict derivations of structural patterns obtained from nature. But even as they sit within this threshold, they concurrently seek to occupy a plane above critical judgment—preferably, Cage invites us to read them as socialized acoustic experiences that are neither ideal nor complete in their identities.

To read Cage's compositional role then as one of activator in, and impresario of, sound-space, may best serve to describe his approach at large, in so much as it was his process of discovering and enabling sounds that encouraged listeners to hear the environmental context as utility for the spatial fluctuation and thematic development of those sources. Nature then was imitated and usurped through sonic materials whose lives and purpose remained simply to exist for the moment as a function of the immediate environmental conditions (whether they be guided or free). This dialectic between music and nature creates sound-spaces that sit at the threshold between the intentional and the non-intentional. Cage is comfortable with such a nexus though as it serves to move music beyond the expectations of the functional—an added consequence of which is the negation of any closely held musical ideals:

"one may give up the desire to control sound, clear his mind of music, and set about discovering means to let sound be themselves rather than vehicles for man-made theories or expression of human sentiments".[34]

Thus Cage's musical sound-space becomes conceptualized in a similar manner to the strict approach of the Taishō garden designer's striving for a melding of traditional Japanese aesthetics and Western notions of 'naturalness.' Cage's sound-space constructions, as musical ecologies, are left then to explore those spatio-temporal dimensions of their own making, and can thus never be truly measured or controlled lest the hand of the designer unnecessarily disrupts the inherent beauty of the unfolding scene.

* * *

Cage's interest in the nexus between music and the natural world as exemplified by the Japanese garden was always to be facilitated through the conduit of a compositional system or compositional process that enabled him to enact a *likeness* to the systems of nature. Cage's *Ryoanji* in itself provides an intriguing example of the importance he places on the compositional framework as an aesthetic device to serve his goals

34 Cage, *Silence*, 10.

for creating a musical ecology, an analogue of nature. What then can be said about the specific qualities of the auditory experience of the Japanese garden? What are the sounds within the garden and how can they be characterized in terms of a natural or designed sound-space? The *kare-sansui* at Ryōan-ji is a landscape whose accompanying sound world is created through elements exterior to the site, which is perhaps a compelling reason in itself to seek out a musical translation, but given the depth of post-Muromachi era compositional approaches to landscape design, what are the evident qualities of nature and the evident considerations of the garden designer that appear within the Japanese garden? Where do these realms intersect and how can these be conceptualized, measured and mapped in a manner that enables further Cage-like translations to proceed? In the following chapters I will be exploring these questions as a response to Cage's compositional precedent of musically translating Ryōan-ji. But rather than exploring these questions purely through the conduit of utilizing musical or artistic processes, I firstly adapt a number of analytical methods. These methods essentially seek to discover the manner in which the sound world of the Japanese garden is one that is constructed through a manipulation of landscape elements that have particular impacts on the acoustic behavior of the garden's sounds and therefore greatly impacts the auditory experience and the spatial semiotics. While Cage's interest in the connection between music and nature is seemingly typified in his work on *Ryoanji*, as a catalyst, it also provides for a number of investigative avenues to emerge that may allow for a closer reading of the sound world of the Japanese garden as a *spatial model.*

Formal Methodologies for Analyzing the Japanese Garden

Two Gardens as Test Cases

Cage's use of the Ryōan-ji garden as a spatial paradigm has become an important artistic precedent, not only in my own investigations, but also for numerous other artists and composers. Though Cage did not delve into the underlying spatial principles of the *kare-sansui* at Ryōan-ji in any formalized manner, his observations of the relationship between empty space and the rocks that are manifest within the garden showed an acute understanding of the aesthetic framework by which the original garden's anonymous designer was by no doubt driven. Later scientific analysis of Ryōan-ji using Medial Axial Transformation (MAT) by Gert Van Tonder and Michael Lyons as well as Semantic Differential Method (SD method) by Miura and Sukemiya have adeptly confirmed that "the hidden structure which observers do not notice affects the various impressions of the garden."[1] As Cage and the many other visitors to the garden over the centuries may also already subtly suggested, Muira and Sukemiya found that the anonymous designer of the rock garden of Ryōan-ji used particular spatial rules and various rhythmic devices to achieve a sense of quietness or stillness to the composition.

This allusion to a sense of balance, stillness and serenity in the composition of the *kare-sansui* at Ryōan-ji represents a basis for the examination in this chapter of what might an investigation into the auditory qualities of the Japanese garden reveal? Whereas Cage used his observations of the physical properties of Japan's most famous *kare-sansui* as the basis for his musical exploration of the space, in effect seeking to *re*-produce the garden as a musical object, the work I will present in this chapter seeks to examine the Japanese garden as an auditory object. Indeed in the case of Ryōan-ji, as Matthieu Casalis observed, there are myriad denotative interpretations of the

---

1 Kayo Miura and Haru Sukemiya, "Visual impression of Japanese Rock Garden (kare-sansui): from the point of view of spatial structure and perspectives cues," in *Proceedings of the International Symposium on ecoTopia Science* (Nagoya: ISETS, 2007), 1168.

garden including the long-held and acoustically suggestive "mountainous islands in a great ocean" or the "tiger with cub leaping over a river."[2] But the idea that the Japanese garden itself may enable particular types of specific auditory experiences to emerge because of its physical design, its materiality, or the sonic content of its interior or its surrounding context has so far been sidelined by a body of research that focuses on the garden as a visual construct laden with hidden compositional or geometric secrets.

In this chapter I will present an investigation into two particular Japanese gardens located in Tokyo, the Taishō-era Kyu Furukawa Teien and Edo period Koishikawa Korakuen, using a range of investigative formal methodologies. The gardens themselves are diverse in size and style, though both are located solely within the historical Japanese gardening tradition and are considered exemplary examples of stroll gardens still extant within the modern-day city of Tokyo. Though they are separated geographically and historically, at their core they share an aesthetic framework for approaching landscape design that was initially imported from Sui Dynasty China during the sixth to eighth Centuries CE and developed and evolved over hundreds of years into a distinctly Japanese artform. I use these two gardens of Kyu Furukawa Teien and Koishikawa Korakuen as test cases, not only as a means to explore the role of acoustic considerations in the experience of the design of the gardens, but also to understand how various methodologies for the analysis of auditory qualities of a space may produce a wide range of insights that enable a deeper understanding between sound and space to emerge. The focus on auditory knowledge of the gardens is facilitated firstly through a general introduction to current discourses within the field of *acoustic ecology*. The discipline was pioneered in Vancouver Canada by R. Murray Schafer and Barry Truax in the late 1960s and is distinguishes itself through a series of mapping and compositional methods that focuses on field-captured audio recordings. I also examine the related theory of *aural architecture* by acoustician Barry Blesser and environmental psychologist Linda-Ruth Salter as a foil to the work of Schafer and Truax. These two theories and their inherent similarities are then drawn on through an application of semiotician Algirdas J. Greimas and his visualization method, or semiotic square, as a means in which to describe the auditory experience of a garden. A final approach to understanding the auditory experience of a Japanese garden is explored through applying the technique of Formal Concept Analysis in which the focus is on the connection between auditory and landscape features of a Japanese garden and how such design features may be represented in abstract and objective terms using the language of mathematics.

2 Matthieu Casalis, "The semiotics of the visible in Japanese rock gardens," *Semiotica* 44/3–4 (1983): 358-359.

Acoustic Ecology

It is perhaps the overwhelming visual impact created by Japanese garden design that has led many recent investigations to attempt to decode the spatial proclivities and underlying properties of such spaces as meaningful manifestations located purely within the concept of landscape as a topographical phenomenon.[3] That the Japanese garden has been most commonly reducible to a tenet of "miniaturization, in which elements such as rocks and ponds are used to represent large-scale landscapes,"[4] is perhaps a natural consequence of the dialectical relationship they establish between what they present as natural and what they presented as artificial. But what is most often overlooked is what impact their spatial manifestations of landscape form have on their resulting acoustic qualities, or what composer and theorist R. Murray Schafer calls the conditions of a *soundscape*.[5]

For Schafer's colleague and co-founder of the discipline of *acoustic ecology*, Barry Truax, a soundscape represents not merely the presence of an acoustic environment (which could be natural or simulated), but also the potential of such an environment to communicate information to a listener.[6] Both Truax and Schafer have argued that, in particular, natural environments and their acoustic behaviors produce particularly meaningful experiences to auditors, and thus the sounds within them constitute a type of mediating language between listener and environment. It is from this position that the field of acoustic ecology arose in the 1960s. Initially conceived as a means to better understand and classify the urban acoustic environment and the plethora of technologically mediated sonic objects within it, Schafer derived the term soundscape from the word landscape. As a foil to his conception that a landscape constitutes all objects within a visible environment,[7] Schafer defines a soundscape as representing all *auditory* phenomena within a given environment. Through the establishment of the World Soundscape Project (WSP) at Simon Fraser University in Vancouver in the early 1970s, a diverse number of researchers from music composition, sociology, acoustic engineering and urban design initiated a multi-disciplinary investigation into urban soundscapes. The group sought to initially investigate the impact of noise pollution within cities as a way to propose new approaches to urban sound design. Indeed Barry Truax had noted that the underlying aspiration of soundscape design is "to discover principles and to develop techniques by which the social,

3 Keita Yamaguchi, Isao Nakajima and Masashi Kawasaki, "The Application of the Surrounding Landform to the Landscape Design in Japanese Gardens," *WSEAS Transactions on Environment and Development* 8/4 (2008): 655-665.

4 David Young, Michiko Young and T. Yew, *The art of the Japanese garden* (Tokyo: Tuttle, 2005), 20

5 R. Murray Schafer, *The soundscape: our sonic environment and the tuning of the world* (Rochester, NY: Destiny Books, 1977).

6 Barry Truax, *Acoustic Communication* (Westport, CT: Ablex Publishing, 2001).

7 Manon Raimbault and Danièle Dubois, "Urban soundscapes: Experiences and knowledge," *Cities* 22/5 (2005): 339-350.

psychological and aesthetic quality of the acoustic environment may be improved."[8] The field of acoustic ecology grew from this research group, as did their codification of a new acoustic terminology and analysis methodology to describe a soundscape and the semiotics of its constituent parts with particular emphasis on the manner in which auditors *relate* to their local sounding environment.

For both Schafer and Truax then, the methodological framework developed for investigations into the acoustic ecology of an environment seeks to be a subtle interrogation of the connections (or disconnections) between auditors and a soundscape. The multi-disciplinary nature of soundscape studies means that numerous methodologies (such as using multi-channel audio recordings, psychoacoustic analysis etc.) become tools for assessing the inherent meaningful relationships that are established between particular *sound classes* and their environmental contexts. Techniques for mapping and documenting sound sources within an environment have traditionally relied on spatial audio recordings, collecting SPL (sound pressure level) data, the generation of auditory diagrams, and more recently, soundscape preference studies such as those developed by Wei Yang and Jian Kang.[9]

But in addition to these, and indeed other more formalized techniques such as auralization and acoustic prediction modeling (both of which Truax refers to as drawing on the traditional *energy transfer model* of analysis), is the innovative case for the utility of *soundscape compositions*. A soundscape composition is a musical work that is pieced together from edited field recordings of human and/or natural environments, and at once seeks to mend the perceived decontextualization that other forms of pure electronic music (such as *acousmatic music*) causes a listener when source sounds become highly abstracted and disconnected from their original forms and contexts. Here, Truax and Schafer, as composers, forward the idea of soundscape composition as a tenant of the *communicational model* of understanding a sounding environment. Soundscape compositions that sample and then remix particular natural acoustic environments thus seek to challenge what Pierre Schaeffer and Jérôme Peignot originally described as the conditions of the *acousmatique* in the 1950s. During this early period of electronic music a new listening context arose in which the audience was confronted with experiencing musical sound signals without performers. Pierre Schaeffer came to define the acousmatique as "referring to a sound that one hears without seeing the causes behind it."[10] Deriving their terminology from the manner in which Pythagoras asked probationary students to listen behind a screen to his lectures as *akousmatikoi*, Schaeffer and Peignot considered loudspeakers as an analogous technological veil that delivered sources disconnected from the immediate

8 "Handbook for Acoustic Ecology," accessed 18 August, 2012, http://www.sfu.ca/~truax/handbook2.html

9 Wei Yang and Jian Kang, "Soundscape and Sound Preferences in Urban Squares: a Case Study in Sheffield," *Journal of Urban Design* 10/1(2005): 61-80.

10 Pierre Schaeffer, *Traité des objets musicaux* (Paris: Le Seuil, 1966), 91.

listening context, a paradigm that Schafer and Truax would later criticize through their labeling of such a condition as *schizophonia*.

In contrast, Schafer, Truax, Hildegard Westerkamp, Darren Copeland and Charles Fox have argued that their soundscape composition have sought to *re-place* the listener back into an audibly meaningful context of identifiable source sounds such as those taken from pristine natural environments. Schafer contends that such an approach is ecological and listener-inclusive because of an avoidance of the unfamiliarity of complex textures, timbres and temporal manipulations—the contexts of the original source sounds become exceptionally important representations within the work. This ecological focus is also manifest in the word soundscape itself, which was a purposeful derivative from the term landscape.

But the classification of sounds within the framework of acoustic ecology also relies on other derivative terminologies: primarily through the use of three fundamental terms used to describe a particular soundscape: *soundmark*, *keynote sound* and *signal*. Like the term soundscape itself, the term soundmark is a derivative from landmark and indicates "a community sound which is unique or possesses qualities which make it specially regarded or noticed by the people in that community."[11] Truax hears church or temple bells, *Adhān* (Muslim call to pray), town square clocks or foghorns as typical examples of soundmarks. Schafer devised the term keynote as one based on the musical notion of key centre or home tonality. It is a means to describe an anchoring sound within a soundscape: "keynote sounds are those which are heard by a particular society continuously or frequently enough to form a background against which other sounds are perceived."[12] Such sounds are also described as drones and examples include the sounds of the sea for maritime communities, air conditioner or fan noise as well as the sounds of traffic in cities. Jean François Augoyard and colleagues[13] consider drones as analogous to a ground against which other acoustic figures emerge, but also as indicators to the qualities of a space in terms of acoustic fidelity. The third key term used for auditory classification is *signal*, and they are regarded as foreground sounds within a soundscape. Within urban contexts this sound class may subsequently be comprised of electronically generated auditory warnings (sirens, horns etc.). In advancing the terminology of *soundmark*, *signal* and *keynote*, Truax and Schafer contend that any sounding environment (human-modified, natural, or virtual) becomes an assessable taxonomy in which information interpreted from the different source classes (as carriers) enables an auditor to construct a phenomological schema.

11 Schafer, *The soundscape*, 10.

12 Barry Truax, *The Handbook for Acoustic Ecology* (Burnaby, B.C: Cambridge Street Publishing, 1999), accessed August 28, 2012, http://www.sfu.ca/~truax/handbook2.html

13 Jean François Augoyard and Henry Torgue, *Sonic experience: a guide to everyday sounds*, trans. Andrea McCarthy and David Paquette (Montreal: McGill-Queen's University Press, 2005).

Though the field of acoustic ecology has been united through this new terminology and has readily embraced the communicational model of analysis forwarded by Truax, Paul Carter has observed a bifurcation of research interests within acoustic ecology because of the diversity of interest from numerous disciplines investigating auditory experience and auditory communication.[14] He sees a split between composers, sound artists and ecological activists in one camp, and historians, anthropologists, acousticians and scientists in another. This seeming diversity has also led Sophie Arkette to critique the fundamental interests common to Truax and Schafer (as composers) as solely reliant on the creative focus of electro-acoustic soundscape composition rather than actual soundscape design interventions within the built environment or soundscape analysis of urban designs. Arkette has noted, that when Schafer declares "the world to resemble an orchestrated composition, he invites us to take assertive action to change its form and content"[15] without understanding or suggesting the greater difficulties of actually implementing sound design into any real-world urban environment.

But for the discipline of acoustic ecology, soundscape compositions themselves are positioned as just one *tool* for ecologically aware listening, though an overwhelmingly favored one. Truax explains that soundscape compositions place prime importance on maintaining the "listener recognizability of the [captured] source material," and that such compositions therefore ought to invoke the listener's "knowledge of the environmental and psychological context." Many of the electronic soundscape compositions of Truax and Hildegard Westerkamp[16] thus function as virtual acoustic experiences in that they *suggest* acoustic spaces that may never be completely formed, built or realized outside of the electronic music studio—space then becomes constructed *through* the sounds, and potentially acquiescent to a preconceived musical narrative. Truax also suggests that for the soundscape composer, "knowledge of the environmental and psychological context must duly influence the shape of the composition at every level." As such, "the work must enhance our understanding of the world and carry its influence into our everyday perceptual habits."[17] But in the same manner as what Terry Daniel and Joanne Vining describe as the initial impetus for environmental activism, the founding concerns of the discipline of acoustic ecology and the role of soundscape compositions were to similarly draw attention to sonic degradation and auditory pollution. As Daniel and Vining describe it:

---

14 Paul Carter, "Auditing Acoustic Ecology," *Soundscape, the journal of acoustic ecology* 4/2 (2003): 12-13.

15 Sophie Arkette, "Sounds like City," *Theory, Culture & Society* 21 (2004): 161.

16 For example see Westerkamp's *Into the Labyrinth* (2002) a work generated from field recordings of urban areas in India, Darren Copeland's *Memory* (1997) from recordings of Stockholm and surrounds, and Truax's *Pacific Fanfare* (1996), a number of sound scenes composed from 10 soundmarks of Vancouver.

17 Truax details the tenets of the soundscape composition manifesto on his website: http://www.sfu.ca/~truax/scomp.html (Accessed January 24, 2013).

"Much of the concern for scenic beauty and other aesthetic or amenity resources grew out of a more general concern for the protection and preservation of the natural environment. The environmental movement was motivated in part by alarm over degradation of the physical/biological environment. Human caused pollution of the air and water, and careless development of land were seen as threatening the natural ecosystem."[18]

This was certainly a shared concern for the acoustic ecology movement, though for Schafer, Truax and others, the notion of pollutants and the degradation of the environment were naturally directed at a concern for new sound sources (particularly electronic ones) and the relatively recent downwards shift in auditory quality of the urban environment. Certainly the focus many acoustic ecologists place on the documentation and preservation of soundscape through audio recordings relegates much of their methodological practice as techno-centric. Because of this, Carter has also argued that the camp of composers, artists and activist's implicit objective is to ameliorate and draw attention to "a neglected dimension of the everyday world, and, by appealing to the listener's musical sensibilities, to enlist support for its preservation and protection."[19] Indeed Schafer's original intentions for soundscape studies were to situate it within,

"a middle ground between science, society and the arts. From acoustics and psychoacoustics we will learn about the physical properties of sound and the way sound is interpreted by the brain. From society we will learn how mankind behaves with sounds and how sounds affect and change this behavior."[20]

But apart then from the use of the studio techniques for generating soundscape compositions, the approach of acoustic ecologists in understanding the boundaries, meanings and function of a soundscape have involved various visual mappings of auditory phenomena as well as soundscape preference studies that apply the key terms of the discipline to ascertain the connection between auditors and their local acoustic environment. This interest in the connection between auditors and their local environment has traditionally been facilitated most commonly through the use of *soundwalking*. Pioneered within the field by Hildegard Westerkamp, soundwalking involves a group of participants following a pre-agreed route through an environment in silence. Westerkamp notes:

18 Terry C. Daniel and Joanne Vining, "Assessment of Landscape Quality," in *Behaviour and the Natural Environment*, eds. Irwin Altman and Joachim F. Wohlwill (New York: Plenum Press 1983), 44.
19 Carter, "Auditing Acoustic Ecology," 12.
20 Schafer, *The Soundscape*, 4.

"A soundwalk is any excursion whose main purpose is listening to the environment. It is exposing our ears to every sound around us no matter where we are. We may be at home, we may be walking across a downtown street, through the park, along on the beach; we may be sitting in a doctor's office, in a hotel lobby, in a bank; we may be shopping in a supermarket, a department store, or a Chinese grocery store; we may be standing at the airport, the train station, the bus stop. Wherever we go we will give our ears priority. They have been neglected by us for a long time and, as a result, we have done little to develop an acoustic environment of good quality".[21]

Interestingly, the similarities between Westerkamp's strategies for soundwalking and Cage's 1972 work *A Demonstration of the Sounds of the Environment* are many, though seemingly approached from opposite ends of the aesthetic spectrum. Whereas Westerkamp uses soundwalking as a tool not only for *auditory awareness* but also in an effort to subjectively assess the quality of a soundscape, the aesthetics of Cage dictate that the notion of a heightened awareness to sounds *reduces* the need for classifications regarding the hierarchical or qualitative judgments of a sounding environment. But for both Cage and Westerkamp, of primary importance is that auditory awareness is always listener-centric such that sounds inhabit a space in which the listener is also within. In presenting such a listener-centric approach, the discipline of acoustic ecology aims at facilitating methodologies that will uncover the structure of exemplar soundscapes (whether they are natural or human-made). When such exemplars are considered as models of design composition, replications and usurpations of their structures may inform better design strategies for future acoustic environments. Indeed, the larger goals of Schafer and the acoustic ecology movement have always been not only to examine the qualities of a soundscape, though to use this effort as a means to ascertain whether there are reoccurring patterns that point towards exemplary soundscape designs. The work of Bernie Krause[22] in bio-acoustics and his *auditory niche theory*, which Truax describes as one in which the natural soundscape is comprised of "sounds uttered by various coexisting species occupying discrete frequency bands that do not overlap,"[23] certainly provides an important pointer towards the composition of pristine natural soundscapes and their qualities as structurally significant. Though for Truax and Schafer, there is an overwhelming and perhaps more pressing desire for a future in which the everyday soundscape of the urban environment becomes readily transformed into an exemplary sound-space. The tools, tactics and techniques of acoustic ecology then are aimed not only as utilities for generating analysis, but equally as catalysts for 'ear cleaning' and implementing large scale urban soundscape upheaval.

21 Hildegard Westerkamp, "Soundwalking," *Sound Heritage* 3/4 (1974): 18-27.
22 Bernie Krause, "Anatomy of a Soundscape: Evolving Perspectives," *Journal of the Audio Engineering Society*, 56/1-2 (2008): 73-101.
23 Truax, *Acoustic Communication*, 82.

The Soundscape of Koishikawa Korakuen

Though there have been numerous investigations within the field of acoustic ecology into natural environments and their soundscape qualities, such as the Japan Society for Acoustic Ecology's *100 soundscapes of Japan* project or the Finnish Society for Acoustic Ecology's *100 Finnish soundscapes*, the examination of traditional Japanese gardens as model environments of auditory design has been somewhat overlooked by the discipline. Through a 2007 field trip project to the Tokyo garden Koishikawa Korakuen, the techniques of Schafer, Truax and Westerkamp were tested in an effort to seek out the possible connection between the underlying structural traits that make the garden at Koishikawa Teien such a recognized exemplar of manipulated landscape form, to its extant soundscape qualities and acoustic design.

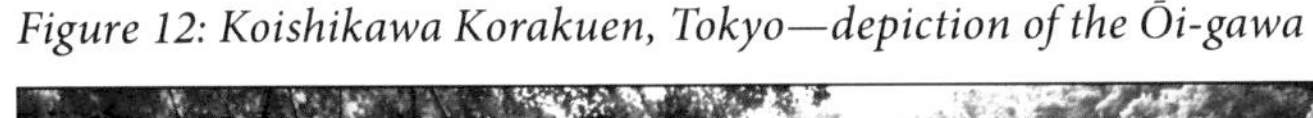

Figure 12: Koishikawa Korakuen, Tokyo—depiction of the Ōi-gawa

The garden dates from the Edo period (1603-1867), and was created for the Mito Tokugawa family by *niwashi* Sahyoe Daitokuji.[24] Located in what is now the heart of Bunkyo Ward in east-central Tokyo, it is a particularly well-known site having been

24 Yoji Aoki, Hitoshi Fujita and Koichiro Aoki, "Measurement and analysis of congestion at the traditional Japanese garden 'Korakuen'," in *Monitoring and Management of Visitor Flows in Recreational and Protected Areas Conference Proceedings*, ed. A. Arnberger, C. Brandenburg, and A. Muhar (Vienna: Bodenkultur University, 2002), 265.

Figure 13: Koishikawa Korakuen—sawatari forming a path along the Daisensui

first surveyed in 1629 by the *daimyō* Tokugawa Yorifusa, and was subsequently finished in its construction by Yorifusa's successor, Mitsukuni. The garden later found the attention of Western criticism by well-known Edo expatriate architect Josiah Conder in 1893.[25]

To the garden's south lies the Kanda River and to the southwest Iidabashi train station while its eastern border is shared with the Tokyo Dome and a large amusement park. Additionally, two major arterial roads that flank its southern and western edges gives the site a sense that it is a stalwart of an almost forgotten history embedded within an ultra modem Japanese metropolis. It can be best described perhaps in the same terms that Marieluise Jonas sees Tokyo's urban interior as one "resisting modernization, it has survived hidden deep in the innermost of the urban giant."[26] But if the garden remains a stalwart, the architectural context of the urban giant is, in the eyes of Botand Bognar, "rapidly modernizing [yet] consistently defying a truly Western-type urbanization."[27] Koishikawa Korakuen is particularly large with an area of 70,847m^2 and consequently contains a number of functioning water features that produce telling soundmarks throughout the site (Figure 14). The large central pond named *Daisensui*, (after a famous Chinese lake originally called *Shifu*), is fed by the waterfall *Shiraito no taki* from a small underground source located at the *Naruto*. In addition to the *Shiraito no taki* there are other active water features including the smaller *Nezame no taki* waterfall to the southeast and the *otowa no taki* water feature to the western boundary. An additional feature is a *shishi-ōdoshi* (or deer scarer) located at the entrance. As a type of water fountain the *shishi-ōdoshi* is comprised of a bamboo tube that is pivoted to one side of its balance point. Water is slowly fed into the tube, which accumulates and eventually causes the tube to rotate at the pivot and expel the water. The heavier end of the tube then falls back against a flat rock, making a sharp percussive sound, after which the cycle again repeats. The device was originally intended to scare wild animals from crops by farmers but found an aesthetic use in the Japanese garden as a punctuating and expressive sound to complement the more constant sounds of falling water elsewhere in the garden.

Yoji Aoki gives the following account of a typical visit to modern day Koishikawa Korakuen:

"Nowadays, people normally enter from the gate of southwest [. . .] They walk at first through an open lawn area and then cross over the west pond. The pond is established as a miniature of Lake Shifu, in the South of China, which is appreciated as a beautiful landscape in China. Then people arrive to a small bridge, called Togetsu Bridge, from where they may look to the famous bank, called Sotei, which is

25 Josiah Conder, *Landscape Gardening in Japan* (New York: Dover, 1964).

26 Marieluise Jonas, "Oku: the notion of interior in Tokyo's urban landscape," in *Urban Interior*, ed. Rochus Hinkel (Baunach: Spurbuchverlag, 2011), 108.

27 Botand Bognar, "Surface above all? American influence on Japanese urban space," in *Transactions, transgressions, transformations: American culture in Western Europe and Japan*, eds. Heide Fehrenbach and Uta G. Poiger (New York: Berghahn Books, 2000), 68.

again a miniature of the original Sotei Bank at lake Shifu. On the other side of the bridge, people find a small waterfall. After that they come up to a mountain, where they may enjoy an overview of the main area of the garden. Continuing the path, they walk up and down a mountain and come across the Engetsu bridge [. . .] They then come up and down a hill and enjoy Iris fields [. . .] After that people stop at Kyuhachiya cottage looking at the main pond on their right side. The fascinating trail then leads the visitors into a clad, beyond which they pass through the ruin of Karamon Gate. They now enter the inner garden, which they may enjoy by a round trip. Back to the gate, they walk to the westward along a narrow trail like in the mountains [. . .] From this area, they also can see the island in the main pond. The island, called Horaijima, symbolizes a kind of paradise. Over a bridge the visitors finally come back to the lawn field at the beginning of the tour".[28]

As is common for an Edo period garden, Koishikawa Korakuen incorporates a number of Chinese influences in its design. As Lorraine Kuck notes, "Chinese touches in Kōraku-en include the Confucian chapel, a Chinese bridge, and the reproduction of Po Chü-i's causeway across the West Lake."[29] Indeed, Mitsukuni sourced the name Korakuen from a Chinese text by Hanchuen called *Gakuyoro-ki* that contains the Chinese teaching: "a governor should worry before people and enjoy after people." This teaching, in which the notion of maintaining a legacy of power and influence for future generations is presented as a virtue, essentially became the design brief for the garden.[30] The Chinese influence also extends into a number of garden ornaments (such as lanterns), plants (lotus), and other landscape typologies depicted within Koishikawa Korakuen Additionally, a number of bridges including the *Tsutenkyo* (half-moon bridge) and *Engetsukyo* (full-moon bridge) in the garden are sourced from influential Ming Dynasty writings (particularly from the work of the Confucian scholar *Shushunsui*). Because of this integration, Koishikawa Korakuen is a particularly well-known synthesis of garden typologies.[31]

But this integration has also extended into its current situation within the ultra-modern architectural context of Tokyo city. Despite this obvious historical juxtaposition within the urban environment, the garden has surprisingly maintained its auditory distinctness, and from within its boundaries appears to maintain a somewhat differentiated soundscape from the immediate external surroundings. This in part is due to the extensive absorbent buffer zone of tree and scrub covered hillock boundaries and the key water features of the garden that provide some auditory masking of exterior sounds.

28 Aoki, Fujita and Aoki, "Measurement and analysis of congestion," 265.
29 Kuck, *The World of the Japanese Gardens*, 241.
30 Toshi Tamura, *Korakuen-shi: history of Korakuen* (Tokyo: Tokoshoin, 1929).
31 Tsutomu Hattori, "Garden Formation of Koishikawa-Korakuen Garden based on 'The Plan of Mitosama Koishikawa Oyashihi's Garden'," *Journal of Agricultural Science* 44/1 (1999): 19-29.

Figure 14: Map of Koishikawa Korakuen

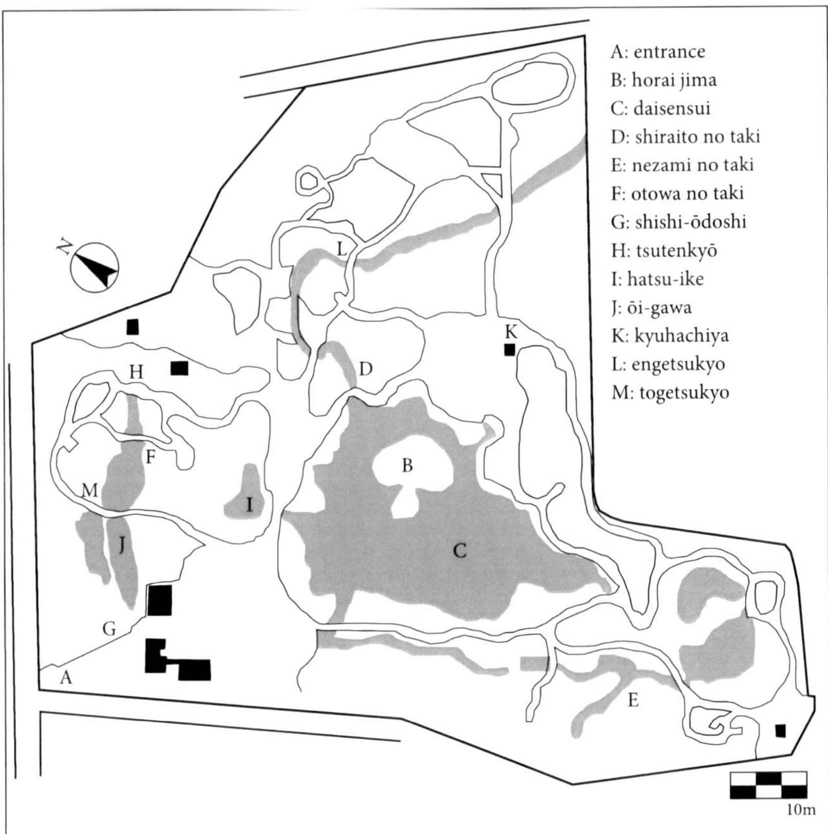

Conder's first Western descriptions in 1893 of the garden are considered yet dogmatic explanations of the topography and plantings which are still current over one hundred years later. As Conder noted too, the garden also utilizes references of known indigenous Japanese rivers such as the *Ōi-gawa* from the Yamashiro prefecture in Japan (Figure 12), as well as a shrine dedicated to the Japanese traveler and poet Saigo. Conder noted other 'exotic' Chinese mythological references as prevalent such as the island of *Horai* (the home of the mischievous spirit creatures *Genii*), and a small lake of white lotus, *Hatsu-ike* (named after a well-known Chinese example). At the time of Conder's first descriptions of the garden, the surrounding area of Tokyo had three million inhabitants. In contrast, the current number of inhabitants of Tokyo has grown to some twelve million people within the twenty-three special wards, which has naturally greatly impacted the acoustic ecology of the city and the garden.

The Spatial Auditory Experience

Though the acoustic ecology of the surroundings of Koishikawa Korakuen have changed considerably since its completion, the nature of the garden's respective design in terms of its soundscape remains unchanged. Though there is much mention in historical Japanese gardening treatises such as the *Senzui narabi ni yagyō no zu* (Illustrations for Designing Mountain, Water and Hillside Field Landscapes), [32] *Tsukiyama teizo den* (Transmissions on Mountain Construction and Garden Making)[33] or the *Sakutei-ki* (Garden Making)[34] on the methods for constructing water features and the need to convey specific scenic effects through site-suggestive topological transformations, no specific details of the actual desired or implied acoustic consequences of these landscape manipulations are mentioned. Nevertheless, the resulting soundscapes produced from such designs often provide evidence of a considered balance between the forces of visual and auditory stimuli. Naturally, the waterfall and its subsequent creek flow is a key sound feature in any Edo era Japanese garden design. Specific tunings are achieved through the placement, size, and shape of rocks within the watercourse as well as the variation in the height of the drop of water, the depth of pond and strength of the water's flow at the waterfall's lip. Such acoustic considerations provide areas within the garden where soundmarks can be designed and differentiated between others within the site. Thus the location of such sites within the larger garden assists in defining the nature of the global sound world that the garden produces. In Figure 15 and Table 1, an initial mapping of sound pressure levels (SPL_A) shows that Koishikawa Korakuen operates through its water features to create areas of auditory interest that are spread throughout the garden's large surface area. These encounters are fostered through the winding paths that take visitors on a circuitous journey. But it is also interesting to note the relationship between the relative distances of encounters as explored in Figure 15.

The distance between the first designed water feature from the entrance to the garden, that is from the front gate to the *shishi-ōdoshi*, when expressed as *a*, can be further assigned to the relative distances between the other designed water features as they occur along the main path from the front gate. As can be deduced from the mapping, there exists mostly an irregularity between the distances to each feature from the path in relation to the first encounter *a*. But the relatively long wait between hearing the small river under the *chūhosen* to the larger *nezame no taki* waterfall is offset by the difference in loudness between the two features.

---

32 David A. Slawson, trans., "Senzui narabi ni yagyō no zu" in *Secret Teachings in the Art of Japanese Gardens* (New York/Tokyo: Kodansha 1987).

33 Einkin Kitamura, *Tsukiyama taizo den*, 1735.

34 Jiro Takei and Marc P. Keane, trans. *Sakuteiki Visions of the Japanese Garden: A Modern Translation of Japan's Gardening Classic* (Boston, Massachusetts: Tuttle Publishing, 2001).

Figure 15: SPL_A mapping of Koishikawa Korakuen

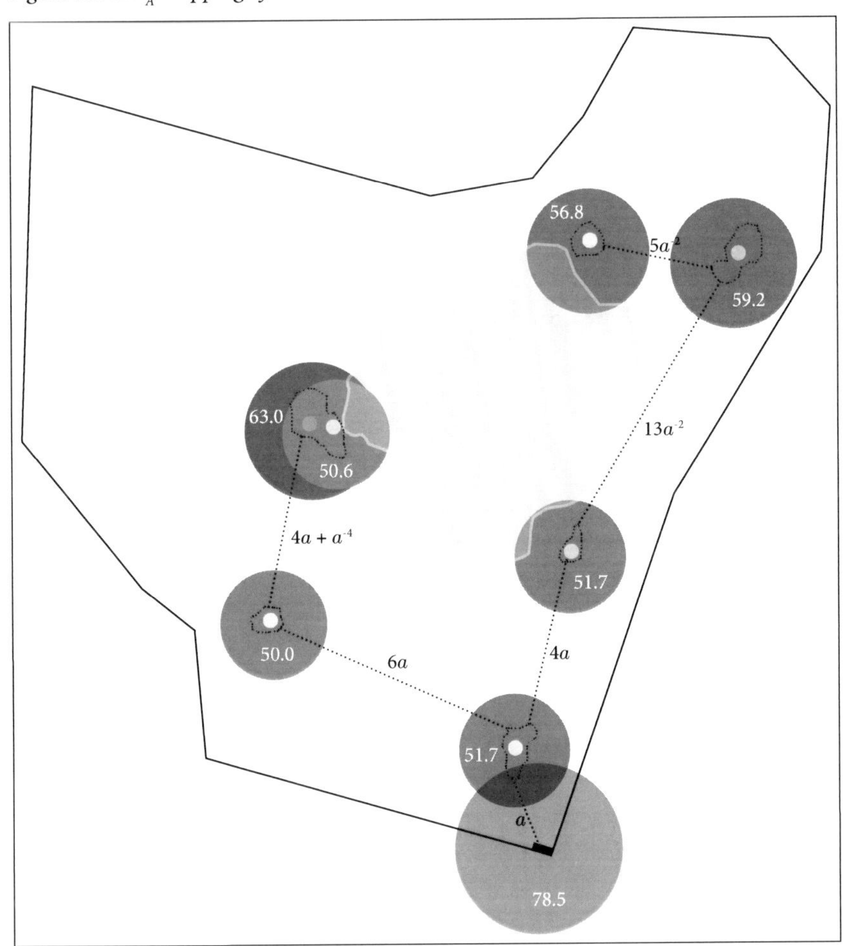

The *nezame no taki* is the main auditory scene of the southern half of the garden and is an aspect that counterbalances the road and traffic noise of the front gate area to that of the rear of the garden where the ambient sound levels are generally more quite. The quietest feature of the garden, the area of the *ōi-gawa* and *otowa no taki* exists as a foil to the more energetic *shiraito no taki*, though the acoustic difference between the features of around 13dB is further emphasized through their relative proximity to each other. Though the distance between the *shishi-ōdoshi* and the *chūhosen* is very close to the distance between the *shiraito no taki* and the *ōi-gawa*, there is a wide difference in the ambient dB levels. Between the former, the levels are equal, and between the latter a noticeable and perceivable difference occurs that would also be amplified by the shorter distance on the path between them. This irregularity between the features plays into the expectations of visitors, in that auditory and landscape features are not presented in a particularly linear or straightforward fashion. Expectations are continually being challenged in the garden by the nature of the paths, topography and strength and intuitive placement of the designed water features.

Table 1: SPL_A Mapping values at Koishikawa Korakuen

Site Name	SPL_A
Entrance	78.5
shishi-ōdoshi	51.7
chūhosen	51.7
nezame no taki	59.2
kisokawa	56.8
shirato no taki	63.0
sawatari (shirato no taki)	50.6
ōi-gawa	50.0

But the localized sounds of designed water features present in Koishikawa Korakuen also reveal both a predictable localized stability (periodicity) yet slight variation among feature sites and their particular soundmarks. They are the most obvious key design features of the garden that provide an aural content which can focus attention locally, and thus away from exterior noise of the city, while at the same time masking these outside noises. The waterfalls thus also articulate a particular region in the garden through an auditory zone or *acoustic horizon*. As Truax explains, the acoustic horizon is:

"The farthest distance in every direction from which sounds may be heard. Incoming sounds from distant sources define the outer limits over which acoustic communication may normally occur, and thus help to define the perceived geographical relationships between communities".[35]

The auditory zones within the garden though are not solely defined by the landscape features such as the paths, plantings or viewing space, but equally by the size, direction and reach of the *acoustic horizon* and the source sounds that acutely define it.

35 Truax, *Handbook for Acoustic Ecology*, 20.

In addition to these considerations, the narrowing of paths, or the slowing down of the traversal through the space via *sawatari* (stepping stones), provides both a visual, as well as an auditory awareness of changed circumstances. Also of note are the loose gravel paths that provide a constant reminder of how human presence shapes the soundscape by providing a soundmark that is a rhythmic interactive foil to the garden's other sounding objects.

Figure 16: Sawatari at otowa no taki, Koishikawa Korakuen

A consequence of these design characteristics is that the symbiotic connection between the visual and the auditory remains in flux. Within Koishikawa Korakuen, and based around its various stopping points, there exist micro environments (or interiors). These environments provide either a designed focal point that is visual and audible (such as a waterfall), or purely visual, though often with an implication of an aural element. For example, a subtle feature of the *otowa no taki* (the waterfall of Otowa-san) at Koishikawa Korakuen is the nature of the randomly placed stones within a shallow expanse of water (Figure 16). Fed by an equally shallow flow of water from an impressive ravine, the *otowa no taki* is an ideal site for bird bathing. In fact, the water is so shallow that the water flow is aurally imperceptible, and the sound sources within this micro environment come wholly from the noises of the wings of the birds and splashing water. Though the name and visual formation of the site suggests a waterfall appearing from a deep ravine, the auditory component arises not through the flow of water, but by the occasional visiting birds interacting with the site.

At Koishikawa Korakuen, stopping points along the paths that envelop the main pond *osensui* provide access to both interior and exterior sensory information, whether it be a self-reflective examination of a waterfall embodying the natural topology of the site, or in the case of the *otowa no taki*, the sound of animal visitors that complement the visual field of play with an sonic object. The dynamics of such garden spaces also comes from the extent of the acoustic horizons, and the subtleties of shift and manipulation facilitated by the design of the garden. For Koishikawa Korakuen, the nature of the topography of each featured landscape is exploited both in the landscape *and* soundscape domain. The large pond of *osensui* duly reflects sounds across the surface, while providing internal sound events created by *koi* (ornamental carp), that unlike the waterfalls or other water sounding devices (such as the *shishi-ōdoshi*) are aperiodic in structure. Thick tree lined groves, and high sculptured absorbing hillocks with between 20-50 meters of green buffer to the main roads act to diffuse sound sources both from within and outside. Each of these features works distinctly in harmony to set a domain for the sonic behaviors possible within the site. They also compress and articulate the larger space into smaller micro environments that are either 'ready-made' with sonic content, or activated by fauna or visitors.

But the garden's soundscape is also assisted in propagating unhindered by the imposing stone walls at the entrance. These barriers work to efficiently reflect the keynotes sounds of the city (low frequency traffic, construction noise, etc) in a particularly marked fashion. Thus the passage from the street into the garden becomes an aurally striking one. The transformation from the soundscape and keynote sounds of the busy exterior city (full of mechanical and electronic sounds) to those of the calming birds, waterfalls and insects, becomes a multi-sensory journey that might also have connotations of traveling back in time. The reduction in loudness at Koishikawa Korakuen from the busy street to the interior of the garden is in the order of +25dB, and occurs rather quickly and within 15m of the street (Figure 15). This transformation in soundscape and introduction of new types of *signals* allows the garden's auditory program to establish itself very quickly. But the garden's *soundmarks* and other *signals* are themselves variable according to the seasons. In summer the *soundmarks* of cicadas and crickets contrasts greatly in terms of their timbre spectrum, frequency bands and rhythmic periodicity to the winter sounds of crows and wind through pine trees. But the soundscape may be also affected through the seasons according to its absorbing surfaces, volume of human traffic and water features whose flows vary throughout the year. The contraction and expansion then of both the auditory content, its qualities of timbre and its reach within the garden gives a flux and dynamic envelope to the soundscape, while still allowing a feeling of encapsulation from the *soto* (outside) world beyond the walls.

Examining the sound world of Koishikawa Korakuen then through the lens of acoustic ecology has highlighted the inherent qualities of sounds within the garden and the relationship of the surrounding context (the *soto*) to that of the interior (or *uchi*). Indeed this particular framework of boundary distinctions has been examined extensively by Merry White, who has argued how notions of inside and outside have pervaded not only the aesthetics of traditional artistic practices in Japan but similarly throughout modern Japanese society. Governing language use and social behaviors the *soto-uchi* paradigm provides a basic axis for defining the relative proximity and intimacy of objects and affiliations to a speaker. They are relational terms that are indexed, and thus fundamentally interconnected through situational context.[36] As inside, "*uchi* evokes the notion of familiarity, proximity, inclusion, certainty and control, [while] *soto* evokes unfamiliarity, distance, exclusion, uncertainty and lack of control."[37] As White contents, these distinctions are not only at play within interpersonal relationships, but within the whole gamut of experience:

"The strictness with which the Japanese boundaries are laid down stems in part from a long cultural tradition of inside-outside distinctions surrounding the use of space and the demarcation of borders [...] Japanese are socialized in early childhood to differences between inside and outside".[38]

The importance then of the sound of water as a signifying presence that evokes feelings of stillness and calmness yet remains also functional in terms of its masking of other *soto* keynote sounds, has highlighted how Daitokuji's design of Koishikawa Korakuen operates through a series of highly considered and balanced encounters in which boundaries between *soto* and *uchi* are subtly marked out or alluded to. These auditory encounters with the water-based soundmarks of the garden are created in such a manner that the entire site is partitioned into smaller (*uchi*) interiors whose acoustic horizons are controlled by both the landscape and soundscape features of the immediate surrounds. But the next question that arises then is how might the semiotics of these water sounds of a Japanese garden be formally investigated and through which methodologies? What is the relationship between landscape formations in a Japanese garden and their soundscape content in regards to traditional Japanese aesthetics? I will be pursuing these questions in the following section through a second case study garden, Kyu Furukawa Teien. I will be applying the semiotic square visualization technique of A. J. Greimas to an example of the construction technique of borrowed scenery or *shakkei* in Kyu Furukawa Teien in an effort to understand the semiotics of the technique as both a visual *and* auditory event.

36 Jane M. Bachnik, "Orchestrated reciprocity: Belief versus practice in Japanese funeral ritual" in *Ceremony and Ritual in Japan, religious practices in an industrialized society*, eds. Jan van Bremen and D. P. Martinez (New York: Routledge, 2003), 113.
37 Haruko Minegishi Cook, *Socializing Identities Through Speech Style: Learners of Japanese as a Foreign Language* (New York: Multilingual Matters Ltd 2008), 43.
38 Merry White, *The Japanese, can they go home again?* (Princeton: Princeton University Press, 1992), 106.

AUDITORY SEMIOTICS

The Technique of *Shakkei*

Though the Japanese language terms for landscape (*fukei* or *keikan*) have been studied in contemporary usages,[39] the concept of soundscape has not gained wide attention. This may be because, as Tadahiko Imada suggests, Schafer's concept of soundscape is Western-centric, though he also notes that "the traditional way of listening in Japan involves a sort of amalgam of environmental sound, instrumental sound and any other environmental facts."[40] Indeed, as I have previously noted, within the three most highly prized and surviving historical Japanese garden treatises there is a concentration on technical and visual aesthetic aspects of garden construction and design, and a stressing on the role of the garden designer as one who synthesizes the extant topographic features of the site with introduced rocks and landscape features so as to suggest naturally found formations and conditions. Though there is no documentation of the potential effects on the surrounding acoustic environment of the introduction of such features, the contemporary Heian text *Genji Monogatari* (Tale of Genji) does contains a somewhat delicate passage on the effects of tuning a water feature:

"The new grand Rokujo mansion was finished [. . .] The hills were high and the lake was most ingeniously designed [. . .] Clear spring water went singing off into the distance, over rocks designed to enhance the music. There was a waterfall, and the whole expanse was a wild profusion of autumn flowers and leaves".[41]

This excerpt suggests that at some level the concept of acoustic design as a gardening technique was evident in the Heian period even though such techniques were never formally codified within later treatises. Indeed the only Japanese gardening technique that has received wide attention, in both the traditional treatises and later discourses outside of the discipline, is the technique of borrowed scenery or *shakkei*.

Kevin Nute[42] describes the technique of *shakkei*, which originated in later Edo period gardens, as one involving the procuring of a remote scene as a way to extend the perceived viewing area of the garden. This is usually achieved through

---

39 Katrin Gehring and Ryo Kohsaka, "'Landscape' in the Japanese language: Conceptual differences and implications for landscape research," *Landscape Research* 32/2(2007): 273-283.

40 Tadahiko Imada, "Acoustic Ecology Considered as a Connotation: Semiotic, Post-Colonial and Educational Views of Soundscape," *Soundscape, the journal of acoustic ecology* 11/2 (2005): 14.

41 E. G. Seidensticker, trans., *The tale of Genji*, vol. 1. (Tokyo: Charles E. Tuttle, 1976), 384.

42 Kevin Nute, *Place, time and being in Japanese architecture* (New York, NY: Routledge, 2003).

the creation of a viewing frame of a low wall or grove of trees or shrubs. The frame trims the raw view aesthetically and may simultaneously incorporate a use of forced perspective as a means to conceal the true distance from the viewer to the exterior landscape feature. As such, Nute observes that,

"in being visually connected to a recognizable feature in the landscape, the viewer not only knows unmistakably *where* they are, but through the apparent merging of the tectonic and the natural, is also made to feel that, like the garden, they too in a sense belong there".[43]

Perhaps one of the most stunning contemporary examples of *shakkei* in Japan occurs in the garden at the Adachi Museum of Art in Yasugi-shi, Shimane-ken (Figure 17). The garden, built in 1970 by Kinsaku Nakane, is located in a favorable topographic context akin to Kyoto's numerous temple gardens enclosed by the Tamba highlands. As David and Michiko Young note, Zenkō Adachi, the founder of the museum, described the garden "as a picture scroll that unfolds before visitors as they stroll through the wings and corridors of the museum, peering through the large plate glass windows."[44]

Figure 17: Adachi Teien, Yasugi-shi

43 Nute, *Place, time and being*, 21.
44 Young and Young, *The Art of the Japanese Garden*, 165.

But the garden is also noteworthy for its use of the technique of *shakkei*, which in a rather virtuosic manner, manages to blur the distinction between what lies inside and outside of the garden. Nakane's use of *shakkei* though goes beyond merely borrowing the exterior landscape, instead he accesses what Teiji Itoh describes as the true meaning of *shakkei*, that of capturing the surrounding landscape hostage:

"In its original sense, however, shakkei means neither a borrowed landscape nor a landscape that has been bought. It means a landscape captured alive. The distinction here is peculiarly Japanese, and it reflects the psychology of the garden designers [. . .] when something is borrowed, it does not matter whether it is living or not, but when something is captured alive, it must invariably remain alive, just as it was before it was captured [. . .] From their [the gardeners'] point of view, every element of the design was a living thing: water, distant mountains, trees, and stones [. . .] Understanding of the term shakkei does not mean a true understanding of the concept unless there is an actual sensation of what it signifies".[45]

In doing so, the garden at Adachi presents the dialectic to the viewer of where does the garden end and the outside begin? From numerous vantage points the foreground garden acts as portal or framing device in which Nakane's attention to design scale, form and local topography, usurp the exterior landscape as synthesized extension of the immediate viewing space. As such, the boundaries of the garden suddenly become infinite, and as Nute has noted, an underlying structural condition emerges that seems to synthesize viewer with landscape.

But the sense of a *shakkei* as operating wholly on the level of the visual articulation of landscape form is not what the original Chinese term exclusively referred to. The term first appeared in the 1634 Chinese garden treatise *Yuanye*[46] and as Wybe Kuitert notes:

"The borrowing of scenery is not only the borrowing of a visual scene: The intent extends to inviting a liberating sense of natural landscape that affects all five senses and differs per season. Nestling swallows are invited, as are soft winds, cool breezes, and the seasonal perfumes of flowers. *Shakkei* must have been understood in this wide meaning in seventeenth-century Japan as well: China and its sense of landscape was no remote thing. Nevertheless it is only the visual aspect of the *shakkei* technique that has recently received much attention".[47]

---

45 Teiji Itoh, *Space and Illusion in the Japanese Garden* (New York: Weatherhill, 1983), 15.
46 Che Bing Chiu, trans., *Yuanje le traite du jardin (1634)* (New York, NY: Besacon, 1997).
47 Wybe Kuitert, *Themes in the history of Japanese garden art* (Honolulu, HI: University of Hawai'i Press, 2002), 177.

Within the garden of Kyu Furukawa Teien, there is also a particularly striking example of *shakkei* that operates within the realm of the garden's soundscape rather than its immediate visual framing of interior and exterior space. By examining the spatial and geomantic context of the garden using the lens of soundscape theory I will seek to extrapolate a deeper picture of the semiotic function of the *shakkei* at Kyu Furukawa Teien using a combination of what Marie-Laure Ryan[48] distinguishes between the coding and diagramming methods of Tzvetan Todorov[49] and A. J. Greimas.[50]

Figure 18: Plan of Kyu Furukawa Teien, Tokyo

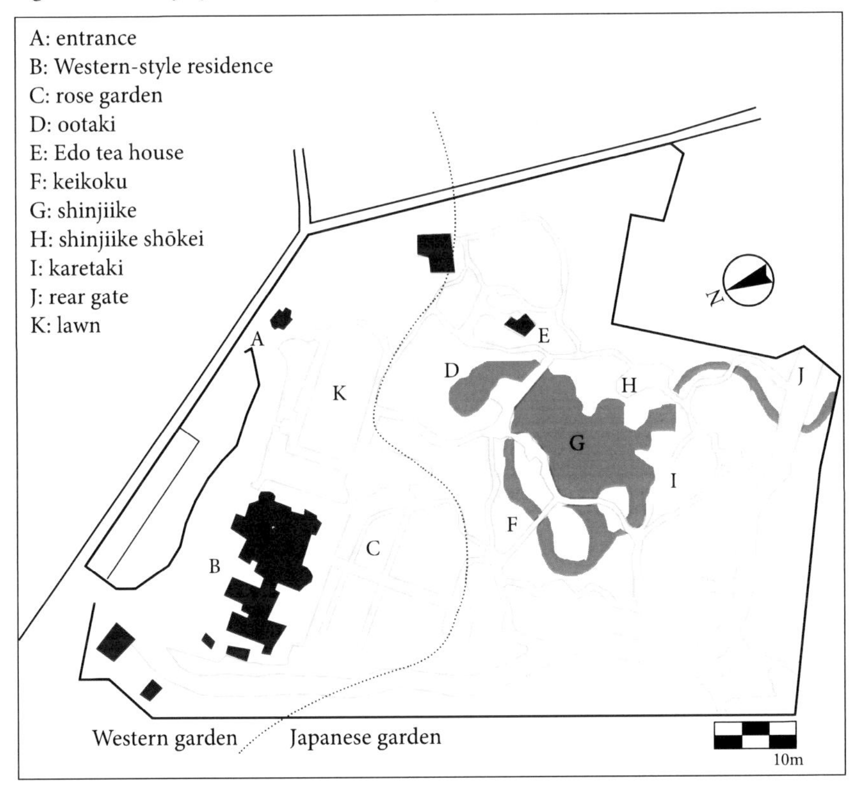

48 Marie-Laure Ryan, "Diagramming narrative," *Semiotica* 165 (2007): 11-40.
49 Tzvetan Todorov, *Grammaire du Décaméron* (The Hague: Mouton, 1969).
50 Algirdas J. Greimas, *Sémantique structurale* (Paris: Larousse, 1966).

The Garden of Kyu Furukawa Teien

Unlike the Edo period Koishikawa Korakuen, the garden of Kyu Furukawa Teien is a later Taishō period garden that was originally completed in 1917 (Figure 18). Located in Tokyo's northern Kita-ku ward it was realized as a compliment to the Western style private residence of the Furukawa family at Komagome. The garden's site is unusual given its favorable and dramatic topographic falloff (around ten meters) from the northeastern to southwestern corners, but also the typically Taishō style rough division of the site into Western influences and Japanese influences. The expatriate architect and *oyatoi gaikokujin* Josiah Conder (1852-1920), who completed an extensive account of the aesthetic landscape traits of Koishikawa Korakuen in 1893, is also connected to Kyu Furukawa Teien. Conder designed the Western style house for the Furukawa family that sits atop the edge of an escarpment, such that with the English lawn, rose bed garden and Western style house, the site is neatly partitioned the from *niwashi* Ogawa Jijei's (1860-1933) Japanese garden below in an synthesis of styles that was common for the period.

But Ogawa's garden itself is also an excellent example of a Taishō period amalgamation of different influences from previous Japanese garden prototypes. Like Koishikawa Korakuen, Kyu Furukawa Teien is a stroll garden or *chisen-shiki-kaiyū-teien* common to the Edo period. A large pond called *shinjiike* is central to the site, from which a large waterfall, *ootaki* (Figure 20), feeds the pond from the northeast and a small river (*suiro*) empties the water to the southwest. Numerous paths circumnavigate *shinjiike* and through the technique of *miegakure* a visitor's movements through the garden is carefully orchestrated so as to both reveal and focus awareness on specific aspects, features, or sounds within the garden at particular points. Numerous paths circumnavigate *shinjiike* and through the technique of *miegakure* a visitor's movements through the garden is carefully orchestrated so as to both reveal and focus awareness on specific aspects, features, or sounds within the garden at particular points. In addition to *shinjiike* and the *ootaki* there are 3 other recognizable (and formally named) landscape features within Kyu Furukawa Teien that are encountered on the circuitous path: *keikoku* (a small ravine), *karetaki* (a dry waterfall) and *shinjiike shōkei* (a seated area for viewing *shinjiike*). Of particular note for this case study is the geomantic connection between the *ootaki* and *karetaki* and the semiotics of the auditory encounter that binds these two features. According to Marc Peter Keane, the use of geomancy (Jp. *eki* or *fūsui*) in Japanese garden design involves the consideration of a "universal structure based on the opposing yet complimentary principles of Yang (the positive, active force) and Yin (the negative, passive force) and their mutual effects on the five basic elements: wood, fire, earth, gold (metal), and water."[51] The influence of Chinese concepts of *Yin* and *Yang*, or in Japanese *in* and *yō*, can be readily traced in the structural relationships within Kyu Furukawa Teien. The most obvious binary pairing within the garden is that of wet and dry, embodied in the *karetaki* (dry waterfall) and its counterpart the *ootaki* (waterfall).

51 Keane, *Japanese garden design*, 24.

Figure 19: Karetaki, Kyu Furukawa Teien

Figure 20: Ootaki, Kyu Furukawa Teien

The *karetaki* (Figure 19) is a feature commonly used in Momoyama era gardens and designed to subtly suggest the movement of water purely through the placement of rocks. A considered articulation of rounded river stones and more oblique textured rock grain is used to represent the falling, flowing and gentle dissipation of water into *shinjiike* from a hidden source. The composition suggests both unique interconnecting scales of time and motion as the illusion of fast flowing water falling downwards is captured in the grain of a large rock in the background of the scene. The moving water is gradually slowed through the intermediate pools below, denoted by small round river stones that mark out the movement of currents and eddies. Eventually the water stalls and spreads as it feeds into *shinjiike*, with large flatter stones or *sawatari* (stepping stones) delicately leading the visitor across this threshold. Ogawa places rounded widely spaced *sawatari* as the means to cross the *karetaki* in an effect that Mara Miller[52] identifies as a commonly used technique for temporal manipulation by Japanese gardeners. The focus and dictation by the garden designer on evoking a slower walking rhythm through the *karetaki* creates a different type of awareness for the visitor that is highly considered and serves a particular auditory purpose. The general slowing down of the visitor at this point allows for what can be argued as favorable conditions for generating a heightened *spatial auditory awareness.*

In the context of Kyu Furukawa Teien, such an opportunity then leads to the revealing of the garden's *shakkei* as one that relies not on a specific capturing of a distance landscape, but a distant soundscape. The situation of the *ootaki* to the north-east working together with the acoustically reflective quality of the water of *shinjiike* and the various bordering rocks of its shallows allows sound to efficiently travel the sixty meters between the two sites, which are nominally hidden in visual terms. This means that at the threshold of the *karetaki* and *shinjiike*, during the visitor's act of negotiating the *sawatari*, one not only sees the denotation of moving water, but also *hears* moving water from the distant *ootaki*. In what Mieko Kawarada and Taiichi Itoh[53] have noted as the inverse relation between the rise in urbanization in Japan and the general decline in visual manifestations of the technique of *shakkei* within Japanese gardens, the encounter at the *karetaki* at Kyu Furukawa Teien validates the indigenous Chinese concept of *shakkei* as encompassing more than simply a visual articulation of space[54]—here it is manifest as an auditory borrowing, facilitated through an auditory window. That the *shakkei* at the *karetaki* is one in which the sound of distant water from the *ootaki* acts as a complimentary element, also points to the geomantic concepts of balancing the forces of *in* and *yō* within the garden.

52 Mara Miller, *The Garden as an art* (New York, NY: State University of New York Press, 1993).

53 Mieko Kawarada and Itoh Taiichi "Environmentalism in Japanese Gardens," in *Environmentalism in Landscape Architecture*, ed. Michael Conan (Washington, D.C: Dumbarton Oaks, 2000), 266.

54 Chiu, *Yuanje le traite du jardin*, 1997.

That these two features are opposite in terms of materiality, the *in* energy within the moving water of the *ootaki* versus the *yō* energy in the static rocks that denote moving water of the *karetaki*, is further offset by their symbolic articulation on the east-west divide of the site.

Diagramming Auditory and Landscape Elements

Examining more closely the auditory environment at Kyu Furukawa Teien, and in particular, the acoustic relationship between the *ootaki* and the *karetaki* reveals an important connection between the auditory objects of *keynote* and *soundmark* and the topographic objects of rocks and water. That there seems to be an inherent interchangeability between Schafer's concept of soundmark and keynote is evidenced when one closely examines the site context and traces auditor expectations of the sounds within that environment. For example, at the *ootaki*, the landscape feature of the waterfall emerging from a ten-meter high drop is complimented by the boisterous soundscape of its falling water. As a telling acoustic phenomenon, those sounds are meaningful within the larger context of the garden and present visitors an articulation of the site in auditory terms. Paths from the garden entrance lead us gradually towards this soundmark, whose situation in the northeast of the garden, and as one emerging from a northern mountain represents what François Berthier[55] describes as a traditionally Daoist conceptualization of space (a common post-Edo influence in garden design). But in an arrangement reminiscent of a traditional Heian-era garden, the energies of *in* and *yō* are balanced in regards to the voids and fluids (falling and still water) countering the masses and solids (rocks and earth).[56]

Walking further along the route and onto *shinjiike shōkei* the visual focus of the *ootaki* is quickly subverted to the qualities of the pond and numerous plantings along the path. After leaving the viewing area of the *ootaki* the sound of the falling water quickly becomes an acoustic background to the qualities of the path and viewing area of *shinjiike shōkei*. The sounds of the *ootaki* have now become a keynote of the garden, a constant background foil propagated by the wide frequency range of the sound of falling water, drop in sound pressure level, filtering effects from trees and ground plantings, and the reflective nature of the rocks and gravel paths. These conditions make the encounter at the *karetaki* directly southwest of the *ootaki* a rather special one.

55 François Berthier, *Reading Zen in the rocks: the Japanese dry landscape garden* (Chicago: University of Chicago Press, 2000).
56 Keane, *Japanese garden design*, 24.

Figure 21: T-function schema, Kyu Furukawa Teien

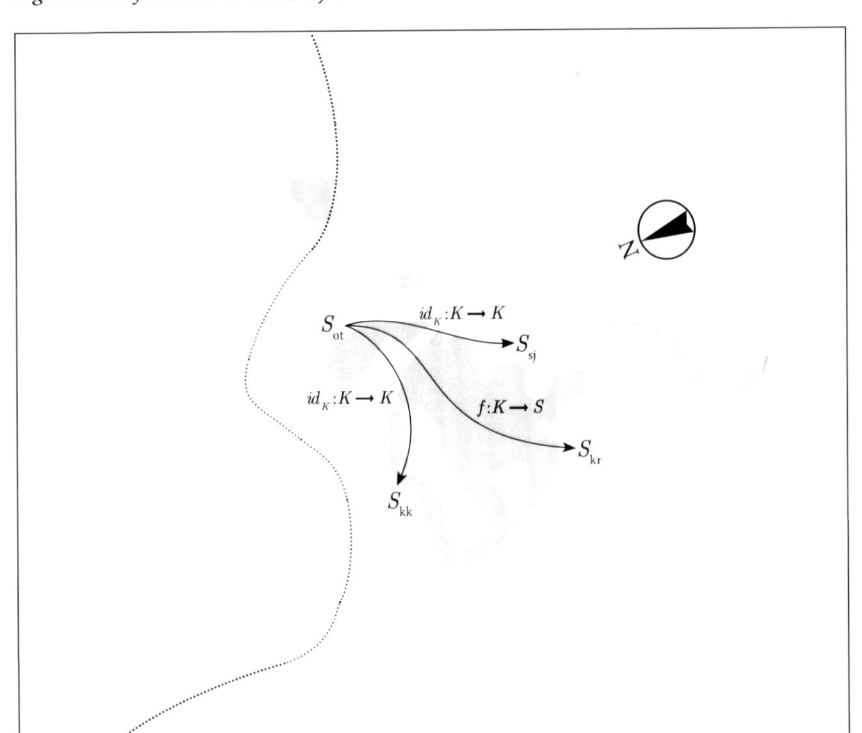

The soundscape at the *karetaki* then is one in which the landscape features readily suggest an auditory analogue. The sounds of the *ootaki*, though in the background and practically functioning as a keynote, are now foregrounded as a meaningful soundmark. The distance sound of falling water is no longer a drone, but has a particular semiotic function: complementing the implied movement of water sculpted in the rocks with an actual acoustic artifact. The *shakkei* at the *karetaki* is then perhaps best conceptualized as one triggered by the recontextualization of the soundmark of the *ootaki*. One can thus consider the connection between landscape and soundscape elements within Kyu Furukawa Teien as one predicated on a sound class transformation function. Given the predominance of the *ootaki* as the primary soundmark of the garden and its acoustic range across the three sites of *keikoku*, *shinjiike shōkei* and *karetaki*, we can readily identify a shared *acoustic arena* in which the sounds of the *ootaki* permeate the site, and thus can be heard from each of the other named landscape features points within the garden (Figure 21). Within the geography of this acoustic arena we can then further deduce 2 types of mappings (or *T*-functions) of soundscape classes of Kyu Furukawa Teien. These occur as both as a mapping of keynote onto itself at *shinjiike shōkei* (S_{sj}) and *keikoku* (S_{kk}), and as a soundmark onto itself at the *ootaki* (operations that can also be described as *identity function* mappings), but also as the transformation of keynote into soundmark at the *karetaki* (S_{sj}). These transformations can be thus expressed symbolically as:

$$T^1 = id_K : K \rightarrow K$$
$$T^2 = f : K \rightarrow S$$
$$T^3 = id_S : S \rightarrow S$$
$$\text{where } f_o id_K = f = id_{Ko} f,\ f_o id_S = f = id_{So} f$$
$$\text{and } K = \text{keynote},\ S = \text{soundmark}$$

The first and third *T*-functions (T^1, T^3) are identity functions which can also be expressed as $f(K) = K$ or $f(S) = S$ where every element of K or S are mapped to themselves. This implies that at the site of the *ootaki*, the emanating soundmark is directly attributable to the immediate landscape feature, and similarly, when at *shinjiike shōkei* and *keikoku* the sound of the distance falling water constitutes a keynote given the immediate landscape conditions at those sites are non-referential in regards to falling water.

The second transformation, T^2, is a function in which a keynote, K, is transformed into a soundmark, S. This is the case at the *karetaki* in which the *shakkei* is one in which the sounds of falling water are denoted in the immediate landscape conditions. This also implies that the acoustic arena (AA) of Kyu Furukawa Teien is expressible as a set of sites for which two important subsets can be defined by the first and second *T*-functions. Using a set builder notation this can be expressed as:

$$\text{Let } AA = \{S_{ot}, S_{sj}, S_{kr}, S_{kk}\},$$
$$AA \supseteq x_{EW} = \{x \in AA \mid \forall x \in AA,\ f(x) = T^1\} = \{S_{sj}, S_{kk}\}.$$
$$AA \supseteq x_{NS} = \{x \in AA \mid \forall x \in AA,\ f(x) = T^2\} = \{S_{ot}, S_{kr}\}.$$

The acoustic arena, AA is a set containing all the named sites of the garden: *shinjiike shōkei* (S_{sj}), *keikoku* (S_{kk}), *ootaki* (S_{ot}), *karetaki* (S_{kr}). When extracting the two subsets, limits corresponding to geographical orientation of east-west (EW) or north-south (NS), allow for derivative subsets that are defined by the two previously described T-functions T^1 and T^2. These two subsets of AA thus reinforce the manner in which garden balances the *in* and *yō* forces given that $\{S_{ot}, S_{kr}\}$ lies on the east-west axis (x_{EW}), and $\{S_{sj}, S_{kk}\}$ lies on the north-south axis (x_{NS}).

Figure 22: Geomantic schema, Kyu Furukawa Teien

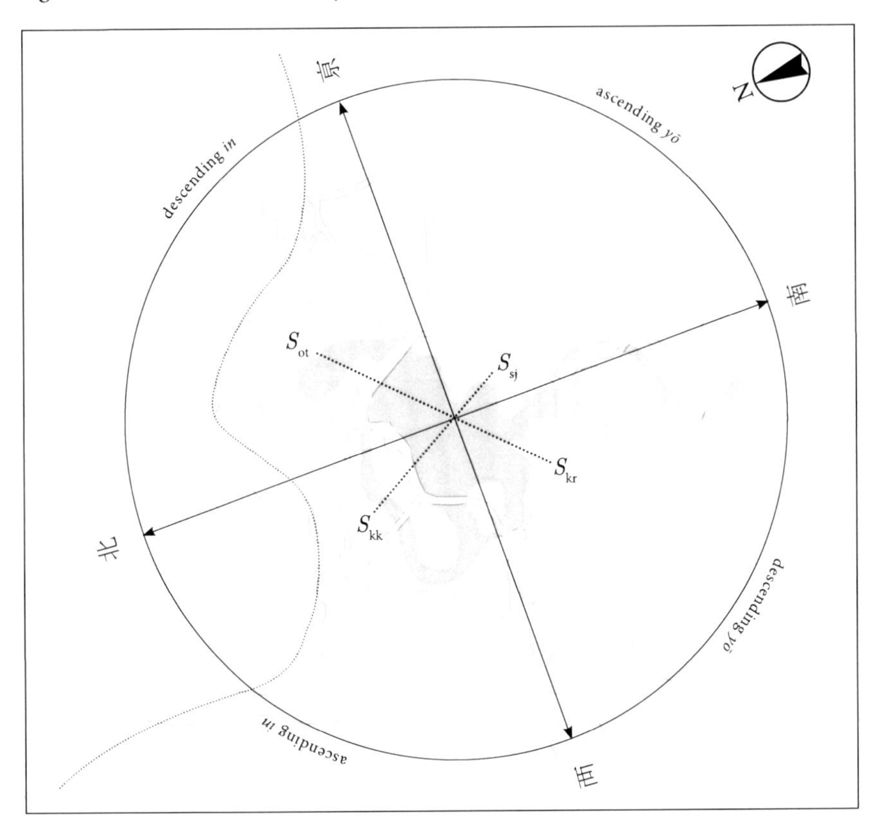

In traditional Heian garden geomancy, the balancing forces at the absolute cardinal points of the compass are also offset by the relationship between quadrants (Figure 22), and also the idea that "almost all manner of things were perceived as being a result of the interrelated effects of the two energy fields [*in* and *yō*]."[57] Thus the pairing of the *ootaki* and *karetaki* in the north-eastern sector and south-western sector also corresponds to the idea that Günter Nitschke[58] identifies as the *topomantic* feature of an idealized landscape which embodies the northern mountain range and the southern fertile plane ('descending *in*' and 'descending *yō*'), and which is also attributable to the pairing of midnight and winter vs. noon and summer. The *keikoku* and *shinjiike shōkei* similarly correspond to the balancing quadrants of 'ascending *yō*' (evening and autumn) and 'ascending *in*' (morning and spring). This concept essentially captures the geomantic dictum that all elements are moving towards their opposites.

57 Jirō Takei and Marc Peter Keane, trans., *Sakuteiki, visions of the Japanese garden* (Boston, MA: Tuttle Publishing, 2000).
58 Nitschke, *Japanese Gardens*, 37.

Such binary pairings of concepts bound to the landscape features of Kyu Furukawa Teien and its corresponding soundscape elements readily allows for a deeper exploration of the *ootaki-karetaki* pair through the construction of a semiotic square (Figure 23). Firstly instigated by structural linguist Algirdas J. Greimas, a semiotic square is a visualization, expansion and development of an initial binary word pair as a means to "explore graphics that formalize theoretical relationships in order to find hidden structures at work in literary narrative."[59] As Louis Hérbert describes, the technique functions through establishing,

"a conceptual network, usually depicted in the form of a square [. . .] [and is] defined as the logical articulation of a given oppositional analysis by increasing the number of analytical classes stemming from a given opposition from two (e.g. life/death) to four – (1) life, (2) death, (3) life and death (the living dead), neither life nor death (angels) – to eight or even ten."[60]

The square is firstly constructed from two terms or concepts named S_1 and S_2 that sit at the square's upper corners and form the basic opposition of the framework. Greimas postulates that "two terms may said to be contrary when the presence of one presupposes the presence of the other, and . . . [the] two terms (S_1 and S_2) are said to be contrary if the negation of one implies the affirmation of the other, and vice versa."[61] From these terms all other elements of the square are derived and thus also the opportunity for interrelationships to emerge. From the establishment of S_1 and S_2, the contradictory terms of $\sim S_1$ and $\sim S_2$ complete the square as its lower corners. Both $\sim S_1$ and $\sim S_2$ can be most simply defined as 'not-S_1' and 'not-S_2,' but these can further become refined to reflect specificities within the narrative text or object under consideration. The final relationships are the *meta-concepts* of which the most prominent are the *neutral term* ($\sim S_1 + \sim S_2$) and the *complex term* ($S_1 + S_2$), but also include the *positive deixes* ($S_1 + \sim S_2$) and the *negative deixes* ($\sim S_1 + S_2$). While the initial S_1 and S_2 pairing for Kyu Furukawa Teien corresponds to the *ootaki* and *karetaki*, I nominate two derivations of landscape elements for $\sim S_1$ and $\sim S_2$. By the account of Hébert, the formal relationship of contradiction between S_1 and $\sim S_1$ is bidirectional, and in the case of the Kyu Furukawa Teien square the terms *sj* (*shinjiike pond*) and *sw* (*sawatari*) are similarly geomantically oppositional. As explored in Figure 23 these contradicto-

59 Algirdas J. Greimas, *Structural Semantics: An Attempt at a Method*, trans. Daniele McDowell, Ronald Schleifer and Alan Velie (Lincoln, Nebraska: University of Nebraska Press, 1983), and Algirdas J. Greimas and F. Rastier, "The Interaction of Semiotic Constraints," *Yale French Studies* 41 (1968): 86-105.

60 Loius Hérbert, *Tools for Text and Image Analysis: An Introduction to Applied Semiotics* (Texto!, 2006), accessed December 22, 2013, http://www.revue-texto.net/Parutions/Livres-E/Hebert_AS/Hebert_Tools.html

61 Algirdas J. Greimas and J. Fontanille, *The Semiotics of Passions. From States of Affairs to States of Feelings*, trans. P. Perron and F. Collins. (Minneapolis: University of Minnesota Press, 1991), 153.

ry terms are derivations from the subsets of AA, x_{EW} and x_{NS}, in which *sj* is located at the intersection between these sets, and *sw* is an element of S_{kr} (*karetaki*).

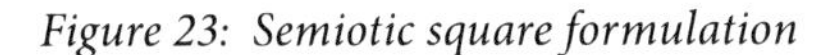

Figure 23: Semiotic square formulation

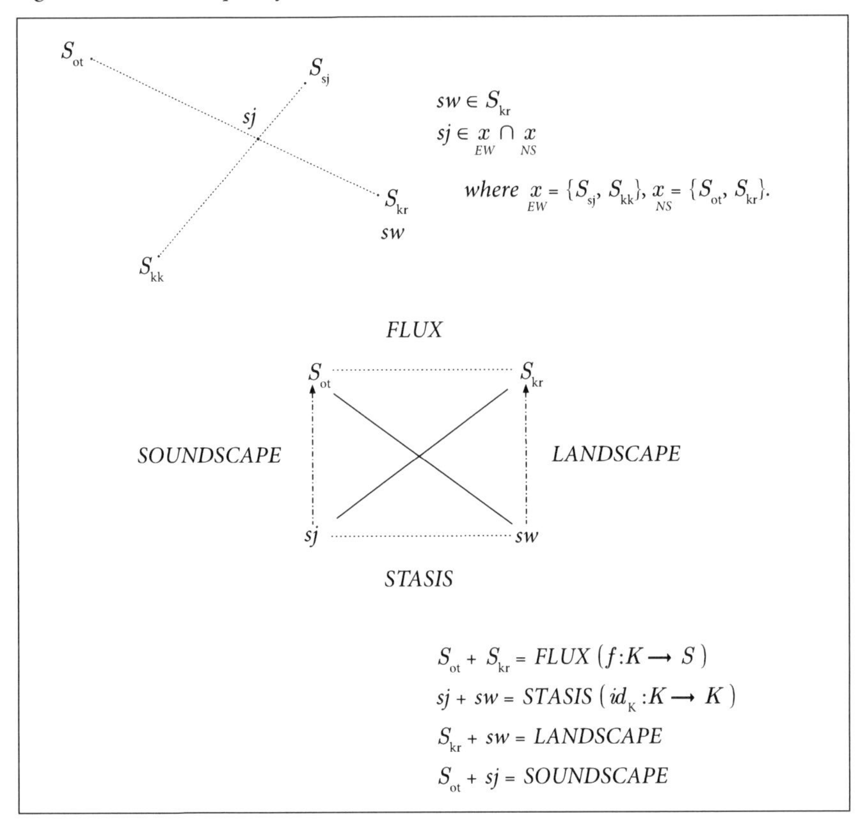

Starting with the implication[62] between $\sim S_2$ and S_1 and Hébert's[63] assertion of the unidirectional nature of this relationship, I have nominated Schafer's term *SOUNDSCAPE* for the positive deixis. As the intensification between the addition of *ootaki* and *shinjiike pond*, the *SOUNDSCAPE* metaterm not only implies the geomantic properties of water as an *in* energy, but similarly the acoustic properties of water within Kyu Furukawa Teien as the primary designed active sound source and facilitator of the reach of the acoustic arena AA. The negative deixis, *LANDSCAPE*, provides a

62 J. Fontanille, *Sémiotique du discours* (Limoges: Presses de l'Université de Limoges, 2003), 60.
63 Louis Hébert, *Dispositifs pour l'analyse des textes et des images* (Limoges: Presses de l'Université de Limoges, 2007).

countering concept in which the *yō* forces of rocks and mountains are embodied in the addition of the *karetaki* and *sawatari*. Finally, the complex and neutral terms capture the T-functions operable between sites within the garden. For the complex term, *FLUX*, the addition of the *ootaki* and the *karetaki* describes the quality of the *shakkei* in the garden as one predicated on the sound class transformation of keynote into soundmark, $f : K \longrightarrow S$. This quality is perhaps the ultimate embodiment of *FLUX*, as a connotation of the unification between the forces of *in* and *yō* in the two sites via a shared acoustic arena, diverse materialities, and the denotation of movement within each site. This sits in contrast to the neutral term, *STASIS*, where I read the addition of *shinjiike* and *sawatari* as analogous to the $f : K \longrightarrow S$ (or id_K) mapping occurring within x_{EW}, for which $sj \in x_{EW}$.

Reflections

To return once again to the concept of *shakkei* and its manifestation at the *karetaki*, one can position the denotation of the landscape within this site as of moving water, though because of the presence of the auditory artifact of the borrowed sounds of the *ootaki*, I would argue that there is a connotation of *balance* that arises. If we examine all the metaterms of the Kyu Furukawa Teien square the concept of the transformation of *in* into *yō* and the geospatial and material combination of these opposing energies within the features of the garden, acts as a directorate to the spatial and semiotic dimensions emerging within the experience of the *shakkei*. The use of rocks as a manifestation of *yō* is countered both by the denotation of representing moving water (*in*) and the acoustic artifact of the *ootaki* soundmark. Added to this, the natural axial weighting between east and west that both features lie on and their mutually opposed yet interdependent material properties work in a manner to harmonize the garden as a series of poised encounters between geomantically disparate elements.

Jijei Ogawa's garden design at Kyu Furukawa Teien then balances both the physical manifestation of LANDSCAPE through highly considered applications of topography, plantings, path structure and accompanying visual encounters *with* the subsequent acoustic behaviors of these elements as a derived SOUNDSCAPE. Similarly, the FLUX and STASIS of the *T*-functions are the connotations of the soundscape elements of keynote and soundmark whose denotations are functionally backgrounded or foregrounded acoustic signals. Though the garden is a product of an early Taishō era industrializing Japan, recently opened up to the outside through the Meiji restoration, the geomantic principles used within the design seem to point back towards well-established techniques of the Heian garden masters. But as Meiko Kawarada and Taiichi Itoh have noted the presence of *shakkei* within an increasingly urbanized Japanese metropolis has necessarily reduced the opportunities for pure visual manifestations of landscape borrowing.[64] It is perhaps that Ogawa identified this condition

64 Kawarada and Taiichi, "Environmentalism in Japanese Gardens," 266.

as an opportunity rather than a limitation, and in doing so enabled a unification of landscape and soundscape elements within the garden. Like the traditional forces of *in* and *yō*, the manifestation of interdependent yet opposite, static yet fluctuating elements of landscape and soundscape provide the experience of Kyu Furukawa Teien as one predicated on encountering a harmonization between the seen *and* the heard.

* * *

This investigation into the relationship between auditory and landscape elements within the context of traditional Chinese/Japanese garden aesthetics has brought to the fore the use of formalized diagramming techniques for the representation of these forces. The seeming effectiveness of this type of approach to pinpoint and contextualize particular characteristics of the garden has raised the question of how much deeper can this type of approach take us towards what Schafer and Truax identified as their initial desire to understand the larger patterns that produce pristine or exemplar soundscape environments? If the underlying goal of the discipline of acoustic ecology is to discover these hidden secrets of proportion and structure in exemplary soundscapes, then perhaps a formalized approach through mathematical modeling may yield further significant advances to achieving their goal. In the following section, I will present an example appropriation method for the representation and abstraction of the spatial predilections of the garden of Kyu Furukawa Teien into mathematical forms, hierarchical relationships and visual diagrams as a direct result of applying the technique of Formal Concept Analysis. I present this case study as yet a further interrogative lens to those presented thus far in the quest to uncover the deeper structural traits of a Japanese garden's design, but to also have these insights represented in a fashion that will allow them to be potentially re-engineered in the future. But in doing so I will firstly introduce the theory of *aural architecture* by Barry Blesser and Linda-Ruth Salter as a means to both strengthen the coverage of discourse on auditory spatiality as well as to provide a depth to the investigation regarding the physical impact of landscape qualities on auditory experience in a Japanese garden.

The Aural Architecture of Kyu Furukawa Teien

In contrast to Schafer's focus on the taxonomy of a soundscape as a distinct and separable entity of an environment or ecology, Barry Blesser and Ruth-Linda Salter have built an acoustic theory more closely aligned with those notions about architectural phenomenology as exemplified by the theorist Alberto Pérez-Gómez.[65] For Blesser

65 Alberto Pérez-Gómez, *Architecture and the Crisis of Modern Science* (Cambridge: MIT Press, 1983).

and Salter the auditory experience of a sounding environment might equally be facilitated through real or virtual means, though their notion of the aural architecture of space is contingent on the materiality, geometry and texture of its context. Using the key terms of *active* and *passive aural embellishment*, an environment's spatial context can be considered not only in regards to Schafer's taxonomy of sound classes, but also the effects of landscape or architectonic forms in filtering, reflecting, diffusing and diffracting sound waves, which consequently create particular acoustic typologies (or identifiable behaviors), and thus particular auditory qualities for a listener:

"Architecture includes aural embellishments in the same way that it includes visual embellishments. For example, a space we encounter might contain water sprouting from a fountain, birds singing in a cage, or wind chimes ringing in a summer breeze—active sound sources functioning as active aural embellishments for that space [...] In contrast, passive aural embellishments, such as interleaved reflecting and absorbing panels that produce spatial aural texture, curved surfaces that focus sounds, or resonant alcoves that emphasize some frequencies over others, create distinct and unusual acoustics by passively influencing incident sounds."[66]

Like Schafer's theory of soundscape, the notion of aural architecture also implies listening as a subjective activity in which particular combinations of active and passive aural embellishments sum to activate an auditor's *spatial auditory awareness*. For Blesser and Salter, and much akin to the theories of Schafer and Truax, aural architecture is a designate of the

"properties of a space that can be *experienced* by listening. An *aural architect*, acting as both artist and social engineer, is therefore someone who selects specific aural attributes of a space based on what is desirable in a particular cultural framework. With skill and knowledge, an aural architect can create a space that induces such feelings as exhilaration, contemplative tranquility, heightened arousal, or a harmonious and mystical connection to the cosmos. An aural architect can create a space that encourages or discourages social cohesion among its inhabitants."[67]

Given then the already established and wide ranging interest in the aesthetics of landscape form in Japanese gardens as places that have also evoked Blesser and Salter's notions of "contemplative tranquility or heightened arousal," the case for considering the greater taxonomy of a Japanese garden as a spatial phenomena—in which objects or elements are sources emerging from landscape *and* soundscape contexts—seems a feasible strategy for analysis. In fact Blesser and Salter have used their theory of aural

66 Barry Blesser and Ruth-Linda Salter, *Spaces speak, are you listening? Experiencing Aural Architecture* (Boston: MIT Press, 2007), 51.
67 Blesser and Salter, *Spaces speak*, 5.

architecture to specifically position the Japanese garden as an example of a highly considered aural design space:

"Japanese garden design, an ancient art form that stylizes and miniaturizes natural environments by creating the illusion of larger ones, includes the aural experience of space. Not only are objects and plants arranged for their visual pattern, but also for their ability to shadow and reflect sound from active sources [. . .] By introducing the aural experience in its design, a Japanese garden becomes the artistic union of a landscape and a soundscape, and its designer a truly multi-sensory architect."[68]

As discussed previously, my own site investigations have conferred with such observations as those by Blesser and Salter. In particular, the presence of a unique multi-sensory *shakkei* at Kyu Furukawa Teien as well as the use of birds sounds to activate the *otowa no taki* at Koishikawa Korakuen as has already been discussed sets these gardens apart from many of their more famous Kyoto contemporaries. That there is more to the underlying design predilections of the original garden makers than that which is satisfied by the visitor's eyes alone, points directly towards the notion of Japanese garden design as indeed a multi-sensory artform. But the typical landscape features of the gardens also combines what Allen Carlson considers the dialectic relationship they construct when elements of the naturally appearing environment (topography, plantings, water) are juxtaposed with artificial architectonic objects (lanterns, bridges, architecture). Indeed all the elements of a Japanese garden, from the topography, rock placements, paths and water features had been already formally codified and named by the fifteenth century. But in the case of Kyu Furukawa Teien, a number of the landscape features are also active aural embellishments and thus contribute to the garden's sound design.

The most prominent of the active aural embellishments of Kyu Furukawa Teien that contribute to the acoustic activation of the site are the *ootaki*, and the *su-iro* (the small watercourse) that empties into *shinjiike* to the southwest. Other sound signals that occur seasonally within Kyu Furukawa Teien include bird life, aquatic animals (such as the movement of ornamental *koi* and turtles) as well as insects (cicadas and crickets during summer months). In contrast, the complementary passive landscape features of the garden act in concert with these active aural embellishments: in particular, the dramatic topography of the escarpment, heavy mature tree plantings at the garden's boundaries, solid high stone walls providing a buffer to the outer streets and the situation of *shinjiike* and its low lying topographic aspect.

68 Blesser and Salter, *Spaces speak*, 66.

Figure 24: Sawatari at karetaki, Kyu Furukawa Teien

Together these landscape elements and their materiality combine with the various acoustically reflective rocks within the site (especially around the formation of the *karetaki*) to enable the sound of the *ootaki* to penetrate the entire garden and produce an *acoustic arena* (a defined yet potentially fluctuating area in which sounds can be heard). In much the same manner as in Koishikawa Korakuen, the landscape and architectonic features at Kyu Furukawa Teien act in such a way as to produce a distinction between the *soto* sounds of the exterior context and of the city, to the quite balanced surrounds of the *uchi* interior. Taking then the theories of both Schafer and Truax and Blesser and Salter as a combinatorial framework for naming and identifying the function of various landscape/soundscape elements may enable a discrete design theory to emerge of the garden. As a particularly relevant question to the discipline of garden design theory, a more generalized larger-scale investigation than was presented in the previous inquiry into Kyu Furukawa Teien's auditory semiotics at the *karetaki* and its striking *shakkei* may allow for a mathematical substantiation of the qualities of landscape ecology that produce or enable particular acoustic behaviors. By positioning the Japanese garden as a *design space* in which a networked structure of related descriptions of partial and/or intentional designs are housed, will allow for a systematic analysis of the design artifacts using the technique of Formal Concept Analysis (FCA). With the technique of FCA, a formal ontology of the garden is constructed in which a systematic investigation into its taxonomy can reveal deeper spatial predilections regarding its design space and therefore those particular spatial predilections that create such unique and well-documented multi-sensory experiences.

Formal Concept Analysis

Generating an Ontology

As Uta Priss notes, FCA has most widely been used in the information sciences, AI (artificial intelligence) research, information retrieval and software engineering as a method for data analysis, knowledge representation and information management.[69] Karl Wolff describes the fundamentals of FCA as a mathematical exploration of incidence relations between objects and attributes,[70] while Frank Vogt and Rudolf Willie define a formal context (as shown in Table 2) as a collection of binary relations between a set of objects, G (derived from the German word *Gegenstände*) and their corresponding attributes M (*Merkmale*).[71] Given that it may be impossible to list all

69 Uta Priss, "Formal concept analysis in information science," in *Annual review of information science and technology* 40 (2006): 521–543.
70 K. E. Wolff, "A First Course in Formal Concept Analysis: how to understand line diagrams," in *SoftStat'93 Advances in Statistical Software* 4 (1994): 429-438.
71 Frank Vogt and Rudolf Willie, "TOSCANA—A graphical tool for analyzing and exploring data," in *Lecture Notes in Computer Sciences 894*, eds. R. Tamassia and I. G. Tollis (Heidelberg: Springer-Verlag, 1995).

attributes of any given object, a specific 'formal' context or closed world is assumed.[72] This formal context, notated as $\mathbb{K}$, is then expressed as: $\mathbb{K} = (G, M, I)$ for which G and M are sets while I is a binary relation between G and M such that $I \subseteq G \times M$. This fundamental tenant[73] then of FCA is understood as object g (where $g \in G$) has the attribute m, and is notated as gIm or $(g, m) \in I$.

Table 2: $\mathbb{K}$ (Kyu Furukawa Teien)

	Rocks	Gardening ornaments	Moving water	External sounds	Topography	Still water	Fauna	Habitatble architecture	Paths	Flora	Earth	Bridges
Sou.			√	√			√					
Arch.		√						√				√
Sea.				√			√			√		
Art.		√		√	√			√	√			√
Nat.	√		√		√	√	√			√	√	
Ter.	√		√		√	√			√	√	√	
PAE	√	√			√	√		√	√	√	√	√
AAE			√	√			√					

AAE = active aural embellishment, PAE = passive aural embellishment, Ter. = terrestrial, Nat. = natural, Sea. = seasonal, Arch. = architectonic, Sou. = soundscape

Table 2 ($\mathbb{K}$) is a generated formal context of objects and attributes from site visit observations of Kyu Furukawa Teien. By utilizing attributes forwarded by Schafer, Blesser and Salter together with Carlson's contention that Japanese gardens are a dialectic construction between 'natural' and 'artificial' elements,[74] the context attempts to canvas qualities of both landscape and acoustic phenomena. But the landscape features of

72 B. A. Davey and H. A. Priestley, *Introduction to Lattices and Order* (Cambridge: Cambridge University Press, 2002).

73 B. Ganter, G. Stumme and R. Willie, *Formal Concept Analysis: foundations and applications* (Springer-Verlag, Berlin, 2005).

74 Allen Carlson, *Aesthetics and the environments, the appreciation of nature, art and architecture* (Routledge, London, 2000).

the garden (Figure 25) also combine in a dialectic relationship due to elements of the natural environment (topography, plantings, water) becoming juxtaposed with artificial architectonic objects (lanterns, bridges, architecture). The additional attributes of *seasonal*, *terrestrial* and *architectonic* similarly provide a scope for interrogating the myriad instances of objects within the garden.

Figure 25: Landscape elements of Kyu Furukawa Teien

Top row (L-R): *keikoku* (ravine), *ishidōrō* (stone lantern), *cha-shitu* (Meiji tea house), *karetaki* (dry waterfall). Second Row (L-R): *sori-bashi* (wooden arch bridge) and *myōseki* (turtle rock), *shinjiike* (pond), *ishidōrō* (stone lantern), *nori-no-ishi* (mountain path). Third Row (L-R): *sawatari* across the *karetaki*, exterior architectural context, *ishibashi* (stone bridge), *tobi-ishi* (stepping stone path). Fourth Row (L-R): *sekitō* (stone pagoda), Furukawa residence, heavy tree cover, *ootaki* (waterfall).

Table 3: Instances of g in G

G	$g \in G$
Rocks	*myōseki* (named rocks), *mumyōseki* (unnamed rocks)
Garden ornaments	*sekitō* (pagoda), *ishidōrō* (stone lanterns)
Moving water	*ootaki* (large waterfall), *suiro* (small water course)
External sounds	traffic, air conditioner noise, aircraft, human activity
Topography	*miyama-no-sakai* (mountainous area), *jiban* (plain or flat ground), *keikoku* (ravine), *karetaki* (dry waterfall)
Still Water	*shinjiike* (pond in shape of the Chinese ideogram for 'heart')
Fauna	*koi* (carp), turtles, birds, insects
Habitable architecture	Furukawa residence, *cha-shitu* (tea house), pavilion
Paths	*tobi-ishi* (stepping stone path), *nori-no-ishi* (mountain path)
Flora	azaleas, conifers, maples, cowberry, honeysuckle, bamboo, plum, cherry, moss, grasses
Earth	soil, sand
Bridges	*ishibashi* (stone bridge), *soribashi* (wooden arched bridge)

But rather than pursuing the impracticality of specifying all instances of garden objects (for example, every species of tree etc.) object classes are used, of which instances (that is, $g \in G$) can be found in Table 3 and represent an exhaustive account of meaningful garden objects. But the power of FCA as a tool for exploring the qualitative dimensions of an ontology is best revealed through a visualization of $\mathbb{K}$. Figure 26 graphs $\mathbb{K}$ of Kyu Furukawa Teien (Table 2) as a partially ordered lattice that Willie[75] notates as $\mathfrak{B}(\mathbb{K})$. This particular lattice was constructed using the software Concept Explorer.[76] As shown, circles denote objects or attributes with labels sitting above indicating an attribute, and those sitting below an object. At the top of the diagram is an empty circle representing the power set, $P(S)$ and at the bottom of the diagram the empty set, $\varnothing$. The *reading rule*[77] of the line diagram states that an object g has an attribute m if and only if there is an upwards-leading path from the circle named by

75 R. Willie, "Restructuring lattice theory: an approach based on hierarchies of concepts," in *Formal Concept Analysis*, ed. Sébastien Ferré and Sebastian Rudolph (Berlin: Springer-Verlag, 2009).

76 S. A. Yevtushenko, "Systems of data analysis 'Concept Explorer'," in *Proceedings of the 7th national conference on Artificial Intelligence* (Moscow: KII-2000), 127-134.

77 Wolff, "A First Course in Formal Concept Analysis," 3.

"g" to the circle named "m". For example, in $\underline{\mathfrak{B}}(\mathbb{K})$ of Kyu Furukawa Teien, the object 'earth' has an attribute called 'Nat.' (Natural).

Figure 26: Line diagram of $\underline{\mathfrak{B}}(\mathbb{K})$ generated using ConExp software

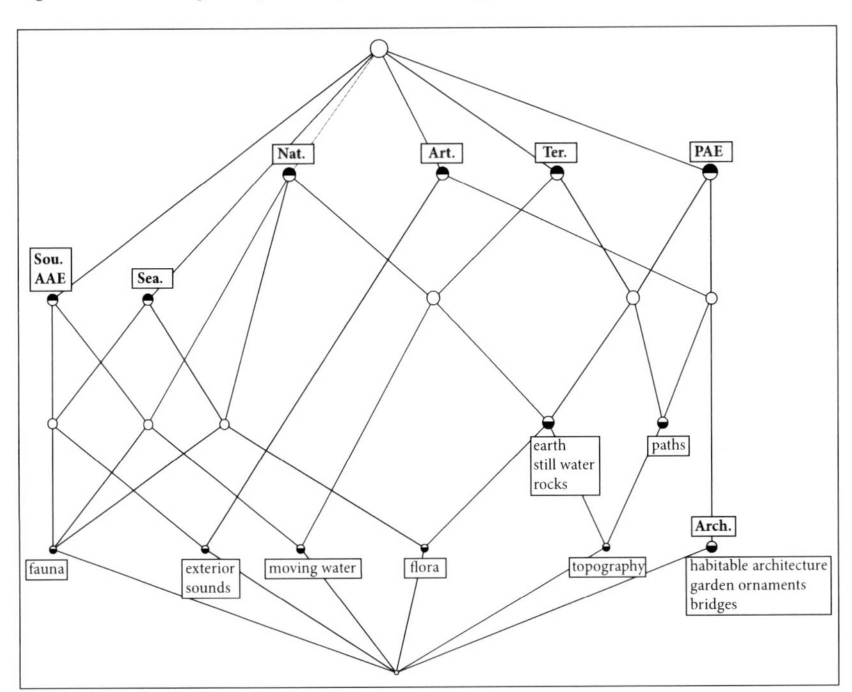

Sou.=soundscape, AAE=acoustic aural embellishment, Sea.=seasonal, Nat.=natural, Art.=artificial, Ter.=terrestrial, PAE=passive, aural embellishment, Arch.=architectonic (See Table 2 for definitions).

Both the power set, and the empty set provide the anchors for the web of relations (or partially ordered subsets) that describe aspects of the garden as an ontology. But within $\underline{\mathfrak{B}}(\mathbb{K})$ of Kyu Furukawa Teien, numerous unnamed nodes also arise, providing a structural glimpse as to the groupings of objects and their interrelations.

From the graph then, what are called *formal concepts* (denoted as $\mathfrak{c}_n$) are derived through a mapping of the intention and extension of a concept node: ext($\mathfrak{c}_n$), int($\mathfrak{c}_n$).[78] The extension of a concept consists of a set of all formal objects, G, that have all the formal attributes, M, of the concept and vice versa.[79] For example, in $\underline{\mathfrak{B}}(\mathbb{K})$ of

78 Priss, "Formal concept analysis," 524.
79 Uta Priss, "Facet-like Structure in Computer Science," *Axiomathes* 18 (2008): 249.

Kyu Furukawa Teien (Figure 26), in the extension of 'Sea.' (Seasonal) are the objects 'fauna' and 'flora,' while in the intention of the object named 'paths' are the attributes 'Art.' (Artificial), 'Ter.' (Terrestrial), and 'PAE' (passive aural embellishment). These subsets of $\mathfrak{B}(\mathbb{K})$ are thus known as formal concepts, for which Peter Burmeister and Richard Holzer[80] use a derivation operator for defining them through assigning arbitrary $X \subseteq G$ and $Y \subseteq M$ conditions that undergo the following mappings:

$$\begin{aligned} X \mapsto X^I &= \{m \in M \mid gIm, \forall g \in X\}, \\ Y \mapsto Y^I &= \{g \in G \mid gIm, \forall m \in Y\}. \end{aligned}$$

As an example, within $\mathfrak{B}(\mathbb{K})$ of Kyu Furukawa Teien, the formal concept seasonal ($\mathfrak{c}_{\text{Sea}}$) is notated as a subset of objects (G) that are within the extension of the attribute 'Seasonal' (Sea.):

$$\{\mathfrak{c}_{\text{Sea}}\} = \left(\{\text{fauna, exterior sounds, flora}\}, \{\text{Sea.}\}\right)$$

But as Priss has noted,[81] the success of a formal context within disciplines outside of those concerned purely with the inherent mathematical qualities lies in the careful modeling of attributes, for which the derived concepts need not specifically correspond to intuitive notions of the user. In Table 3 a detailed definition of M is presented along with examples of G for $\mathbb{K}$ of Kyu Furukawa Teien.

Given that in FCA a formal concept is understood as a pair of subsets of $\mathfrak{B}(\mathbb{K})$ in which $(A, B) = \mathfrak{c}_n$, with $A \subseteq G$, and $B \subseteq M$, such that $A = B^I$, and $B = A^I$, Ganter and colleagues note that this condition leads to further hierarchies to emerge.[82] Of primary interest is what they describe as the *subconcept-superconcept-relation* within $\mathfrak{B}(\mathbb{K})$, which is operable under the following conditions:

$$(A_1, B_1) \leq (A_2, B_2) :\Longleftrightarrow A_1 \subseteq A_2 \,(\Longleftrightarrow B_1 \supseteq B_2).$$

Here, concepts that sit higher up within the lattice, and contain themselves lower-level concepts within their extension are known as *superconcepts* and are notated as $(A_2,$

80 Peter Burmeister and Richard Holzer, "Treating incomplete knowledge in Formal Concept Analysis," in *Formal Concept Analysis: foundations and applications* (Berlin: Springer-Verlag, 2005), 115.

81 Priss, "Formal concept analysis in information science," 532.

82 Ganter, Stumme and Willie, *Formal Concept Analysis*, 84.

B_2). These superconcepts may include lower level attributes, or *subconcepts*, (A_1, B_1), which then correspondingly account for the notion of an inheritance of attributes or class inclusion.[83]

Table 4: Instances of g in M in $\underline{\mathfrak{B}}(\mathbb{K})$ with respect to G

M	Definition of *M*	*G*
Soundscape	Sound class instances (signals, soundmarks, keynotes) according to Schafer's (1977) theory of soundscape	Fauna, moving water, exterior sounds
Active aural embellishment	Blesser and Salter's (2007) definition of sound sources of an environment as agents that provide acoustic identity to a space	Fauna, moving water, exterior sounds
Seasonal	Objects or agents that are influenced by seasonal changes	Flora, fauna, exterior sounds
Natural	pertaining to the natural environment	Rocks, moving water, topography, flora, earth, fauna, still water
Artificial	Objects that are produced or designed or serve functional purposes: also ephemera that are a function of human activity	Garden ornaments, exterior sounds, topography, habitable architecture, paths, bridges
Terrestrial	Features or land-form elements of the earth	Topography, flora, moving water, still water, paths, earth, rocks
Passive aural embellishment	Blesser and Salter's (2007) definition of the acoustic properties (materiality) of a space in governing the behavior of sound sources (filtering, diffraction, reflection, absorption)	Topography, flora, still water, paths, earth, rocks
Architectonic	Pertaining to the built environment or architectural features or fabrication	Habitable architecture, garden ornaments, bridges

83 Priss, "Facet-like Structure in Computer Science," 250.

But another interesting property that arises from a partially ordered lattice are unforeseen concepts such as those ones unnamed in $\mathfrak{B}(\mathbb{K})$–see Figure 26. The one instance of a formally named subconcept, $\{\mathfrak{c}_{\text{Arch}}\}$, architectonic, within the lattice occurs as a subset of the extents of the superconcepts $\{\mathfrak{c}_{\text{Art}}\}$ (artificial), and $\{\mathfrak{c}_{\text{PAE}}\}$ (passive aural embellishment). By tracing the intent of the set of objects that intersect these concepts, we find:

$$\{\mathfrak{c}_{\text{Arch}}\} = \left(\left\{ \begin{array}{c} \text{habitable architecture,} \\ \text{garden ornaments,} \\ \text{bridges} \end{array} \right\}, \{\text{Arch., Art., PAE.}\} \right).$$

$$\text{ext}(\mathfrak{c}_{\text{Arch}}) = \left\{ \begin{array}{c} \text{habitable architecture,} \\ \text{garden ornaments,} \\ \text{bridges} \end{array} \right\}$$

$$\therefore \{\mathfrak{c}_{\text{Arch}}\} \leq \{\mathfrak{c}_{\text{Art}}\}, \{\mathfrak{c}_{\text{PAE}}\}$$

From this example, what Ganter et al. describe as the *attribute implications*[84] can be deduced under these conditions:

$$A \rightarrow B \Leftrightarrow A_I \subseteq B_I \;\; \textit{such that} \;\; A, B \subseteq M$$

Which in $\mathfrak{B}(\mathbb{K})$ of Kyu Furukawa Teien corresponds to the notion that {Arch.}→{Art., PAE} (that is, the attribute architectonic implies the attributes artificial *and* passive aural embellishment).

Design Praxis and Ecology

One of the more striking characteristics that emerges from $\mathfrak{B}(\mathbb{K})$ of Kyu Furukawa Teien is the sense of a distinct partitioning between the sets of attributes {Sou., AAE, Sea., Nat.} and {Art., Ter., PAE, Arch.}. Using ConExp software for line diagram generation,[85] the node sizes of the attributes are varied according to object extension count (larger the diameter the greater number of objects within the extent). The six unnamed subconcepts within the diagram highlight its hierarchical partitioning, though assigning an arbitrary naming convention allows a deeper interrogation into the nature of the partitioning within $\mathfrak{B}(\mathbb{K})$. If assigning the three far left subconcepts

84 Ganter, Stumme and Willie, "Formal Concept Analysis," 120.
85 S. A. Yevtushenko, J. Tane, T. B. Kaiser, S. Objedkov, J. H. Correia and H. Reppe, *ConExp* software available at http://conexp.sourforge.net.

as c_α, c_β, c_γ then the extents of these subconcepts and their implications are comprised of the following:

$$\begin{aligned}\text{ext}(c_\alpha) &= \{\text{fauna, exterior sounds}\},\\ \{c_\alpha\} &\rightarrow \{\text{Sou., AAE., Sea.}\}.\end{aligned}$$

$$\begin{aligned}\text{ext}(c_\beta) &= \{\text{fauna, moving water}\},\\ \{c_\beta\} &\rightarrow \{\text{Sou., AAE., Nat.}\}.\end{aligned}$$

One of the first readily understood characteristics to emerge from this sub-lattice is of the attribute equivalence relation between Schafer's notion of {Sou.} and Blesser and Salter's concept of {PAE}, which can be identified through the subconcept-superconcept relations of c_α and c_β:

$$\begin{aligned}&\{c_\alpha\} \le \{c_{\text{Sou}}\}, \{c_{\text{AAE}}\}.\\ &\{c_\beta\} \le \{c_{\text{Sou}}\}, \{c_{\text{AAE}}\}.\\ &\quad \textit{and}\ \text{ext}(\{c_{\text{Sou}}\}) = \text{ext}(\{c_{\text{AAE}}\})\\ &\quad\quad \therefore\ \{\text{Sou}\} \equiv \{\text{AAE}\}\end{aligned}$$

Further to this is the association between each of the attributes of this sub-lattice. The union of these three subconcepts highlights a feature of the Kyu Furukawa Teien sub-lattice that might best be described as one concerned with, or pertaining to the *Ecological* (as a landscape *and* soundscape phenomena) given the nature of the objects as living or dynamic organic instances that may produce active acoustic phenomena. By considering the instances, a and b in the extension ($a \in A$), and intention ($b \in B$), of the summed subconcepts, $\bigcup\mathfrak{C}$, reveals:

$$\begin{aligned}&\bigcup\mathfrak{C} = \{c_\alpha \cup c_\beta \cup c_\gamma\},\\ &a \in \bigcup\mathfrak{C} \iff (\exists A \in \mathfrak{C}, a \in A),\\ &b \in \bigcup\mathfrak{C} \iff (\exists B \in \mathfrak{C}, b \in B),\end{aligned}$$

thus:

$$\text{ext}(\{c_\alpha, c_\beta, c_\gamma\}) = \bigcup_{A\in\mathfrak{C}} A = \left\{\begin{array}{l}\text{flora, fauna,}\\ \text{moving water,}\\ \text{exterior sounds}\end{array}\right\}.$$

$$\text{int}(\{c_\alpha, c_\beta, c_\gamma\}) = \bigcup_{B\in\mathfrak{C}} B = \left\{\begin{array}{l}\text{Sou., AAE,}\\ \text{Sea., Nat.}\end{array}\right\}.$$

That Schafer and Truax have positioned soundscape studies as a form of ecological analysis of site despite criticism from Tim Ingold[86] and Paul Carter[87] that their approach lacks a greater engagement with the environment as a totality remains a point of contention, though within $\underline{\mathfrak{B}}(\mathbb{K})$, the suggestion that Kyu Furukawa Teien's ecological elements might be described in terms of qualities relating to acoustic *and* environmental systems points toward a multi-sensory dimension of the garden. What I have interpreted here as pertaining to the *Ecological* stems from the inheritance of the attributes of $\bigcup\mathfrak{E}$—*soundscape, active aural embellishment, seasonal, natural*—whose corresponding objects represent a distinct subgroup within $\underline{\mathfrak{B}}(\mathbb{K})$. Similarly, the inclusion of the object 'exterior sounds' highlights Blesser and Salter's notion that the acoustic arena of an aural architecture may extend well beyond the visual horizon. Its inclusion here alludes to the fact that considerations about the ecology of the garden must regard the greater context of the site beyond it own demarcations. But by the same token, this concept may also be applied locally within the garden when reading it as a series of connected landscape encounters. Through the traditional technique of *miegakure*, a visitor's movements through the garden is carefully orchestrated so as to both reveal and focus awareness on specific aspects, features, or sounds within the garden at particular points. This could then account for the previous study of *shakkei* operating at the *karetaki*. Here, the "outside" sounds are captured from beyond the local visual horizon (yet still within the greater confines of the garden) and usurped to provide an auditory signal for the obvious denotation of moving water encoded in the *karetaki*'s sculptural rock formation.

But that the sub-lattice of $\bigcup\mathfrak{E}$ can be viewed as a descriptor of the notion of the *Ecological* is further alluded to when considering the intersection between the extent of the three subconcepts:

$$\bigcap\mathfrak{E} = \{\mathrm{ext}(\mathfrak{c}_\alpha) \cap \mathrm{ext}(\mathfrak{c}_\beta) \cap \mathrm{ext}(\mathfrak{c}_\gamma)\},$$
$$a \in \bigcap\mathfrak{E} \leftrightarrow (\exists A \in \mathfrak{E}, a \in A),$$

$$\textit{thus: } \bigcap_{A\in\mathfrak{E}} A = \{\text{fauna}\}.$$

The object {fauna} then is perhaps the axiomatic element of this sub-lattice of $\underline{\mathfrak{B}}(\mathbb{K})$, and moreover both an active acoustic phenomenon of the garden's soundscape ecology as well as a crucial element in its landscape ecology. That it may function as both a design element within the garden (e.g., as with those introduced animals such as ornamental *koi* and turtles), as well as a representative symbol of the natural ecosystem (e.g., through migratory birds, insects) implies that the garden is a highly dynamic

86 Tim Ingold, "Against Soundscape," in *Autumn Leaves: sound and the environment in artist practice*, ed. A. Carlyle (Paris: Double-Entendre, 2009), 10-13.
87 Carter, "Auditing Acoustic Ecology," 12–13.

design space. This also suggests that the garden is bounded by the intersection between those obvious manifestations in which the designer is wholly present, and those in which nature seems to dominate.

But that the object "earth" is vacant from $\bigcup\mathfrak{C}$ is the next point of investigation for the secondary group of unnamed subconcepts. Naming the secondary group of subconcepts (left-to-right), $\mathfrak{c}_\delta$, $\mathfrak{c}_\varepsilon$, $\mathfrak{c}_\zeta$, the (A, B) values and implications are comprised of:

$$\left(\left\{\begin{array}{c}\text{moving water, earth,}\\ \text{still water, rocks,}\\ \text{topography}\end{array}\right\}, \{\mathfrak{c}_\delta\}\right) \; \textit{and}: \{\mathfrak{c}_\delta\} \rightarrow \{\text{Nat., Ter.}\}.$$

$$\left(\left\{\begin{array}{c}\text{earth, still water,}\\ \text{rocks, flora,}\\ \text{paths, topography}\end{array}\right\}, \{\mathfrak{c}_\epsilon\}\right) \; \textit{and}: \{\mathfrak{c}_\epsilon\} \rightarrow \{\text{Ter., PAE}\}.$$

$$\left(\left\{\begin{array}{c}\text{paths, topography,}\\ \text{architecture,}\\ \text{garden ornaments,}\\ \text{bridges}\end{array}\right\}, \{\mathfrak{c}_\zeta\}\right) \; \textit{and}: \{\mathfrak{c}_\zeta\} \rightarrow \{\text{Art., PAE}\}.$$

$$\textit{thus } \{\mathfrak{c}_{\text{Arch}}\} \leq \{\mathfrak{c}_\zeta\}.$$

I consider this area of the Kyu Furukawa Teien lattice as suggestive of the notion of *Design Praxis* if one considers this collection of formal objects as associated with the parameters and tools of garden design. The objects and spatial typologies located in this partition of the lattice have been named, codified and well documented in the traditional Japanese garden treatises and represent a repertory of compositional forms and structures available to the garden designer. That they include elements of a garden that may produce not only visual articulations of site but auditory behaviors points towards the notion that Japanese garden design contains potential structures for synthesizing manifestations of landscape and soundscape form. Examining more closely the secondary group of subconcepts as a united group, $\bigcup\mathfrak{D}$, the extent and intent contain pertinent qualities and objects that are distinct from $\bigcup\mathfrak{C}$. Defining the conditions for finding and the instances, b in the intention $(b \in B)$, and instances of a in the extension $(a \in A)$, the objects and attributes of $\bigcup\mathfrak{D}$ are:

$$\begin{aligned}
&\bigcup\mathfrak{D} = \{\mathfrak{c}_\delta \cup \mathfrak{c}_\epsilon \cup \mathfrak{c}_\zeta\},\\
&a \in \bigcup\mathfrak{C} \Longleftrightarrow (\exists A \in \mathfrak{D}, a \in A),\\
&b \in \bigcup\mathfrak{C} \Longleftrightarrow (\exists B \in \mathfrak{D}, b \in B),
\end{aligned}$$

thus:

$$\text{ext}(\{\mathfrak{c}_\delta, \mathfrak{c}_\epsilon, \mathfrak{c}_\zeta\}) = \bigcup_{A\in\mathfrak{D}} A = \left\{\begin{array}{c}\text{moving water, flora,}\\ \text{earth, still water, rocks}\\ \text{topography, paths}\\ \text{architecture, garden ornaments,}\\ \text{bridges}\end{array}\right\},$$

$$\text{int}(\{\mathfrak{c}_\delta, \mathfrak{c}_\epsilon, \mathfrak{c}_\zeta\}) = \bigcup_{B\in\mathfrak{D}} B = \left\{\begin{array}{c}\text{Nat., Art.,}\\ \text{Ter., PAE.}\end{array}\right\}.$$

That the superconcepts $\{\mathfrak{c}_{\text{Nat}}\}$, $\{\mathfrak{c}_{\text{Art}}\}$, $\{\mathfrak{c}_{\text{Ter}}\}$, $\{\mathfrak{c}_{\text{PAE}}\}$ represent elements that inform $\bigcup\mathfrak{D}$ reveals that the notion of the design of a Japanese garden involves a *balancing* of qualities in terms of acoustic behaviors and the nexus between natural and artificial landscapes. Though the complimentary attribute to *passive aural embellishment* is located within $\bigcup\mathfrak{E}$ (i.e., *active aural embellishment*), the common object 'moving water' acts as a pivotal element between each sub-lattice of $\mathfrak{B}(\mathbb{K})$. This then means that as an active aural embellishment, 'moving water' is an object that is both an element for landscape/soundscape design as well as a constituent of an ecological system. But the complete extents of $\bigcup\mathfrak{D}$ contains a further common object, 'flora' and one shared attribute, *natural*, which seems to further suggest that the structure of $\mathfrak{B}(\mathbb{K})$ implicates the design space of Kyu Furukawa Teien as one that convalesces notions of the *Ecological* with *Design Praxis*:

$$\bigcup_{A\in\mathfrak{E}} A \cap \bigcup_{A\in\mathfrak{D}} A = \{\text{flora, moving water}\}.$$

$$\bigcup_{B\in\mathfrak{E}} B \cap \bigcup_{B\in\mathfrak{D}} B = \{\text{Nat.}\}.$$

$$\textit{where}: \{\text{flora}\} \in \{\mathfrak{c}_\epsilon\} \longrightarrow \{\text{Nat.}\},$$

$$\textit{and} \quad : \{\text{moving water}\} \in \{\mathfrak{c}_\delta\} \longrightarrow \{\text{Nat.}\},$$

That the lattice might be read in this fashion allows for a generalized design theory to emerge from the formal context of Kyu Furukawa Teien. Using the premise that the extent of attributes can be understood as a synthesis between active notions of the *Ecological* interleaved with considerations about *Design Praxis* allows for a proposition that the garden is a highly considered entity that seeks to balance what Carlson identifies as manifestations between the artificial and the natural in a Japanese garden. The distinctiveness of the sub-lattices yet their shared elements is further revealed in $\bigcup\mathfrak{D}$ whose single element "topography" acts as a compliment to the object "fauna" which represents the nexus of $\bigcup\mathfrak{E}$:

$$\bigcap\mathfrak{D} = \{\mathrm{ext}(\mathfrak{c}_\delta) \cap \mathrm{ext}(\mathfrak{c}_\epsilon) \cap \mathrm{ext}(\mathfrak{c}_\zeta)\},$$

$$\left(a \in \bigcap\mathfrak{D}\right) \longleftrightarrow \left(\forall A \in \mathfrak{D}, a \in A\right),$$

$$\bigcap_{A\in\mathfrak{D}} A = \{\text{topography}\}, \bigcap_{A\in\mathfrak{D}} A = \{\text{fauna}\},$$

$$\left\{\bigcap_{A\in\mathfrak{D}} A, \bigcap_{A\in\mathfrak{D}} A\right\} \in \mathrm{ext}(\mathfrak{c}_{\mathrm{Nat}})$$

The situation of the objects {fauna}($\in\bigcup\mathfrak{E}$) and {topography} ($\in\bigcup\mathfrak{D}$) and their shared attribute *natural*, alludes to the distinct yet interleaved nature of Kyu Furukawa Teien's design space. That the study of what Zeev Naveh and Arthur Lieberman[88] identify as the nexus between landscape design and landscape ecology has only recently been extended by considerations concerning acoustic phenomena[89] has been a point of inquiry central to this investigation. What is evident then at least within this cursory study of $\underline{\mathfrak{B}}(\mathbb{K})$ is the that objects that relate to the landscape ecosystem and soundscape ecosystem are located within the ext($\bigcup\mathfrak{E}$), and their implications point towards the attributes {Sou., AAE, Sea., Nat.}. In terms of garden design, the lattice indicates that those objects that contribute to its geospatial attributes, ext($\bigcup\mathfrak{D}$) are variously acoustic {Sou., AAE, PAE}, architectonic {Arch.} or a landscape feature {Art., Ter., Nat.}. That the object 'earth' is contained in the partition $\bigcup\mathfrak{D}$ rather than within $\bigcup\mathfrak{E}$ is a function of the object's intent, which includes: {Nat., Ter., PAE}. If 'earth' had been designated {*seasonal*}, it would of been located differently, and though it shares the attribute {Nat.}$\in\bigcup\mathfrak{E}$, this characteristic perhaps highlights what Priss[90] identifies as the utilitarian nature of FCA to visualize the assertions at the point that the lattice was made in terms of the definitions of M in $\mathbb{K}$.

Towards a Design Theory

As a method for the examination of the taxonomy of Kyu Furukawa Teien, FCA has allowed for a deeper level of investigation regarding the garden's design space. In reading the formal ontology of Kyu Furukawa Teien as an amalgam between the *Ecolo-*

88 Zeev Naveh and Arthur S. Lieberman, *Landscape ecology: theory and application* (Springer-Verlag, New York, 1984).
89 See in particular Barry Truax and G. W. Barrett, "Soundscape in a context of acoustic and landscape ecology," *Landscape Ecology* 26 (2011): 1201–1207, M. G. Turner, "Landscape ecology: the effects of patterns on progress," *Annual Review of Ecological Systems* 20 (1989): 171–197, I. S. Zonneveld, "Scope and concepts of landscape ecology as an emerging science," in *Changing landscape: an ecological perspective*, eds. I. S. Zonneveld and R. T. T. Forman (New York: Springer-Verlag, 1990), and Z. Naveh and A. S. Lieberman, *Landscape ecology: theory and application* (New York: Springer-Verlag, 1984).
90 Priss, "Facet-like Structure in Computer Science," 250.

gical and *Design Praxis* creates what I believe is a new form of analysis relevant for design theory. That design theory (particularly within architecture) has commonly focused around analysis using philosophical modes of inquiry,[91] semiotics[92] or shape grammar[93] points to the diversity and complexity of the field of inquiry. What has been generally lacking in the discourse though is an investigation into the application of phenomenological acoustic theories, or how such theories might be systematically appropriated to varied design contexts or extant modes of inquiry. Perhaps an advantage of FCA for design theory, and one revealed in the formal context of Kyu Furukawa Teien, is the systematic nature of the construction, reading and interpretation of the lattice whose principles are based on mathematical relationships, yet are a function of the observed world. That a design might be understood in terms of its underlying visual and auditory attributes, which then are used in the construction of formal concepts is a powerful property of FCA. Similarly, the notion that an object within an ontology can be described as an element of a larger collection (a garden, building or design), and whose relationship within this collection can be accounted for in a very specific non-technical way enables a high degree of malleability in the communication and dissemination of the structure to other designers and design theorists. In this sense, FCA may be a valuable utility that not only describes designs, but aids in their construction.

In the most general terms, this study into Kyu Furukawa Teien has also revealed that though the documentation of Japanese garden design techniques concerning acoustic qualities are historically scant, there are inherent instances within the garden, which due to their attribute qualities, synthesize to create particular landscape/soundscape encounters. The significance of the intersection between the sub-lattices of $\mathfrak{B}(\mathbb{K})$, through the attribute *Natural* and objects {flora, moving water} perhaps encapsulates the deeper purpose of a Japanese garden to emulate the natural environment. But for the garden designer, the manipulation and integration of moving water and flora have particular acoustic consequences that may richly inform the resulting design space. By delicately balancing the impact between landscape form and soundscape stimuli, the notion that a garden is a multi-sensory environment contained within a landscape ecology presents the garden designer with a spatial framework to operate within. This is perhaps most readily revealed in Kyu Furukawa Teien in the use of *shakkei* where the intersection between the Natural, Ecological and the artificial (*Design Praxis*) produces a particularly considered landscape/soundscape encounter at the *karetaki* that at once emulates the serene topography of a natural environment, as well as the controlled hand of the intervening garden designer.

91 K. Michael Hays, *Architecture Theory since 1968* (MIT Press, Cambridge, 2000).
92 See Jeffrey Broadbent, "Recent developments in architectural semiotics," *Semiotica* 101/1-2 (1994): 73–101, and C. Dreyer, "The crisis of representation in contemporary architecture," *Semiotica* 143 (2003): 163–183.
93 W. J. Mitchell, *The Logic of Architecture: Design, Computation, and Cognition* (Cambridge: MIT Press, 1990).

Landscape-Soundscape Synthesis

The formal examination of the *shakkei* at Kyu Furukawa Teien and the larger investigation into the synthesis between notions of sound design and notions of landscape design also has implications for my initial study of Koishikawa Korakuen and the conditions presented at the *otowa no taki*. Recall that a subtle feature of the *otowa no taki* is the nature of the randomly placed stones or *sawatari* within a shallow expanse of water that flows out of a 15m deep ravine. Though the formal name suggests a waterfall, it is implied that we are located well down stream given the classic yet radical artificial scaling (or forced perspective) of landscape form from the ravine which lays only about 10m away, yet seems like it may be 10km away given the rapid change in the surrounding landscape topography. This sense of scale is emphasized by the extremely shallow flow of water from the ravine—making it therefore an ideal site for bird bathing.

In fact, the water is so shallow that much like the auditory illusion of water flow at the *karetaki* at Kyu Furukawa Teien, the flow here at the *otowa no taki* is aurally imperceptible. There is a denotation of moving water, though because of the radically adjusted scale in relation to the viewer, there is no actual referential soundmark or auditory connotation of water movement. What does emerge from Sahyoe Daitokuji's garden design then is another form of *shakkei*, though not the typical Edo period visual manifestation of a distant mountain or architectural feature. In fact the borrowing arises from the visiting bird life. The sounds that do emerge come wholly from the chattering birds and the noises from their wings splashing in the water. But the formal naming of the site also points towards this subtle design feature through a distinct play-on-words. The name *otowa* is both a Japanese family name (hence *otowa no taki*—the waterfall of Otowa) as well as a play on the meaning of the word *wa*, which can also mean feather. Together with the word *oto*, which means sound, the site *otowa no taki* might also be read as the "waterfall of the sound of feathers."

In this case, the *otowa no taki* provides the infrastructure for a sounding environment to emerge through an interaction with the fauna of the garden: in a sense the space is an instrument that is activated through occupation. The soundmarks of the birds are borrowed in this *shakkei* setting but only because Daitokuji's design enables and encourages this condition to flourish. Though the name and visual formation of the site suggests a waterfall appearing from a deep ravine, the auditory component arises not through the flow of water, but by the occasional visiting small birds interacting with the site. This is in contrast, though no less of an exemplar of landscape-soundscape synthesis, as the suggestion of water at the *karetaki* in Kyu Furukawa Teien. There, water is presented as a spatial allusion in which the designed garden, as fixed entity, is crafted to represent a dynamic movement of natural elements. The coalescing of these elements completes Jijei Ogawa's equally subtle examination on the multi-sensory potential of a garden to synthesize tangible and intangible phenomena. As acute listeners walk across the *karetaki* they are accompanied by the sounds of distant falling water from the *ootaki*. The distant source, hidden from view in what

Schafer and Truax would identify as an *acousmatique* condition by way of the screen of topography and plantings, is physically represented in a local spatial manifestation within the container of the *karetaki*. As a *shakkei* the relationship between the *ootaki* and the *karetaki* unites the senses and completes the pairing between *yin* and *yang*—dry and wet, action and non-action. Here, as at the *otowa no taki*, elements of the garden itself are captured alive—the tangible and intangible qualities between *soto* and *uchi* become formed as a singular entity in which the apparent silence of the *karetaki* or the *otowa no taki* is offset by the power of the *soto* sound sources as a welcomed acoustic interlopers.

These scenes of *shakkei* encounters in Koishikawa Korakuen and Kyu Furukawa Teien creates a paradigm for returning to Cage, and in particular, to re-consider the nature of the composer's most infamous musical work *4'33"*. Often cited as a particularly notorious composition, the work is scored for any number of performers who sit in silence for a pre-determined yet chance-derived amount of time (of which the title reflects this time bracket). All exterior or extraneous sounds that emanate from the audience or outside the concert hall constitute the thematic content of the piece. Given the first performance by pianist David Tudor at the Woodstock Music Festival in 1952 was realized for a time bracket of four minutes and thirty-three seconds, the work has subsequently appropriated this as its name, though strictly speaking, the duration is to be chance determined by the performer(s) before the work is to be performed. As a piece that is ostensively about silence then, *4'33"* presents to the listener a unique circumstance in which unintentionally through non-action becomes a framework aided by site context as thematic provider and enabler. The work establishes the same focused opportunity for *shakkei* as the encounters at the *karetaki* at Kyu Furukawa Teien and the *otowa no taki* at Koishikawa Korakuen when one considers that the listening mode of *4'33"* is equally framed in a container discretely nominated by Cage as latent with auditory expectation—the traditional concert hall.

Tudor's first performance inevitably casts the work as a dismantler of contemporary perceptions of exactly which auditory objects belong within music-space. But by considering the work as functioning like a *shakkei* allows notions about the intimate connection between site and music to be further revealed. In *4'33"* those same paradigmatic spatial conditions of container latent with auditory expectation are played out in the equally design controlled environment of the concert hall. Though the materiality and topography of Kyu Furukawa Teien and Koishikawa Korakuen is keenly observed and subtly manipulated by Ogawa and Daitokuji as a means to position the garden as a mirror to the natural landscape, Cage's intentions are not dissimilar. In composing by the dictum that music should represent nature in all her manner of operations places Cage in a position to question the architectonic context and functionality of the traditional concert hall. Naturally occurring sounds, or ones not codified by the institution of the concert stage now provide a rich repository for expectant transformation. For James Peterson, this institutional relationship between traditional site and traditional musical materials situates Cage's compositional methodologies as accessing an "art-process schema" in which "any material organized in any way could have aesthetic value, and traditional aesthetic forms needlessly restricted the range of options open to the artist".[94] As such, reading the container of Tudor's first performance of *4'33"*, and its expectant soundscape experience, would mean the auditor's relationship to the site context naturally embodied an anticipatory listening mode pregnant with musical expectation. Audience members, using the pointers suggested to them by the architectonic articulation, programmatic overlay and material occlusion of the site, came with expectations of experiencing soundmarks. As an acoustic community, an audience's collective concert experience of soundmarks as unique, and specially regarded set aside those keynotes of the *soto* (urban) landscape beyond as unexceptional and only occasional interlopers rather than permanent residents of music-space. Cage's demand for hearing keynote sounds in *4'33"* as *soundmarks* manages to fold the architectural context in on itself. Within the stoic interior of traditional spaces for music listening, Cage's seemingly ascetic stance focuses the notion of keynotes as ones internally generated (unintentionally) by the audience. Thus like the container of the *karetaki* and the *otowa no taki*, Cage's local spatial directive becomes an enabler for the permitted (or perhaps necessary) transformation of these keynotes into soundmarks.

This attention then to site and role of the designer of a Japanese garden as one that coalesces the natural and artificial is played out within the Japanese garden design through particular presentations of landscape form and viewer expectations as to the meaning of these forms. Japanese garden design uses specific and *direct* landscape manipulations to elicit particular soundscape encounters. These encounters are most obvious through the use of water or water sounds emanating from features. The fascination with large ponds, water courses and waterfalls confirms for visitors a soundscape built around these fundamental sounds, which range not only from the sound of water hitting water or water hitting rocks, but also the sound of *koi* rising, turtles, small creeks or rapids and the *shishi-ōdoshi*—that which can be attributable to the notion of the ecological. What is significant though in the focus on such a family of sounds and their landscape forms that generate them are the other effects and indeed the potential of a garden designer using their notion of design praxis to skew or manipulate expectations of the visitor. Thus there is a number of *indirect* ways in which auditory encounters in the Japanese garden arise as a result of the interplay between notions of the ecological—as guided by nature, and design praxis—as instigated by the garden designer. Perhaps the most forceful, as I have already discussed is *shakkei*. But working equally with this technique are the artificial hillocks and morphing of topography which may in many instances not only serve to create the illusion of distant mountains or a valleys through forced perspective and manicured scrubs, but can equally assist in lowering sound levels of the exterior through filtering, sound absorption and diffraction. At the most subtle level perhaps is the materiality of the paths, bridges and *sawatari* where in former times, the sounds of *geta* or wooden

94 James Peterson, *Dreams of Chaos, Visions of Order: Understanding the American Avant-garde Cinema* (Detroit: Wayne State University Press, 1994), 77.

shoes would have a particularly telling effect between encounters with gravel, wood, granite or earth. Such changes in the delicate sounds that the visitor enacts within the garden would not of been lost on even late Meiji era culture, with these differences most likely considered as implying more than a simple route through the garden, but more likely a route through the *time* of the garden.

This is a concept that was apparent to Michel Foucault and is suggested at within his notion of the *heterotopia*: a place of juxtaposed times, a transitional space or disputed territory where activities or knowledge exists outside of the world. Essentially a heterotopia is a space of otherness and one that is necessarily comprised of many layers. As such, Foucault sees the traditional Oriental garden as wholly embodying the third principle of what defines a heterotopia such that:

"The heterotopia is capable of juxtaposing in a single real place several spaces, several sites that are in themselves incompatible. Thus it is that the theater brings onto the rectangle of the stage, [. . .] but perhaps the oldest example of these heterotopias that take the form of contradictory sites is the garden. We must not forget that in the Orient the garden, an astonishing creation that is now a thousand years old, had very deep and seemingly superimposed meanings [. . . .] The garden is the smallest parcel of the world and then it is the totality of the world. The garden has been a sort of happy, universalizing heterotopia since the beginnings of antiquity."[95]

Foucault's notion that a heterotopia is a force within a space that produces alternative narratives through layers of meaning is particularly true of the Japanese garden given its proclivity towards the juxtaposition of landform references (such as the inclusion of famous Chinese peaks such as Mt. Horai), ancient and modern epochs and exotic plantings (which may cover references to landscape typologies of India, China and Japan). As Foucault explains further:

"Heterotopias are most often linked to slices in time - which is to say that they open onto what might be termed, for the sake of symmetry, heterochronies. The heterotopia begins to function at full capacity when men arrive at a sort of absolute break with their traditional time."[96]

These temporal and landscape juxtapositions in the Japanese garden not only serve a traditional didactic means for erudition (such as the referencing of famous or foreign landmarks, architecture, mountains or rivers), but also assists in creating an otherness to the space, which in turn enables even those circuitous journeys within the garden to take on mystical dimensions via travels to other lands and other epochs.

95 Michel Foucault, "Of Other Spaces, Heterotopias," *Architecture, Mouvement, Continuité* 5 (1984): 48.
96 Foucault, "Of Other Spaces, Heterotopias," 49.

Thus the Japanese garden is as much a space intended for didactic encounters as it is one designed for creating purely aesthetic or pleasure-driven ones.

Hence, those sounds that accompany a visitor's journey, or indeed the sounds that are present within each garden scene, play an important part in constructing the deeper narrative of the garden. Of course other mobile fauna add to this scene making and perhaps most pertinently, present a potential for seasonal diversity within the garden. This is achieved through particular planting regimes that according to tree or shrub species enable acoustic arenas to emerge according to where birds are feeding, roosting and congregating, or cicadas and crickets are singing. The connection then between where particular trees or flowers are located greatly impacts what the possible soundscape qualities will be at various times in the year and no doubt will have influenced the placement and planning of the Japanese garden and its sense of a placemaker for otherness.

* * *

If this chapter has been primarily concerned with better understanding the elements of a Japanese garden, the potential meanings of these elements and the qualities of the auditory encounters that are a functions of these elements, then in the most general terms, it can be said that the Japanese garden does indeed represent a multi-sensory *spatial model.* In the case of the Edo garden of Koishikawa Korakuen and Taishō garden Kyu Furukawa Teien, the design of each space has been planned in such a fashion that opportunities to accentuate not only encounters with those obvious and expected landscape elements such as hillocks, paths, valleys and ravines are evident, but also a vast taxonomy of auditory encounters that range from the sounds of falling water, to bird, insect and aquatic life: all of which are temporally connected to the ephemerality of the seasons. What has been perhaps traditionally overlooked then in the Japanese garden is the *connectedness* of the landscape to its acoustic properties, and those particular auditory encounters that it encourages, maintains and offers up to visitors. Indeed, as Blesser and Salter have noted, all manipulations of the physical environment, whether it is within the confines of an interior space such as a bedroom, or within an exterior urban context such as a park, have particular consequences for the behaviors of sounds, and therefore our experience of the space itself. These may of course be manipulations that are too trivial to create a noticeable or perceivable difference to our auditory perception—the movement of a chair or furniture, the planting of a low hedge—nonetheless the notion is an exemplary one. What the Japanese garden suggests then is that those design strategies that have enabled it to evolve over the course of some ten centuries have essentially removed the notion of triviality in regards to the direct relationship between physical form and its auditory consequences. Manipulations in landscape form, shape, materiality and distance are combined in a virtuosic fashion as a series of relationships that could be in its most simplistic terms, described as a parametric model. Whether the designer is completely aware of the exact relationships that define the model is unclear from the historical treatises, though the evidence that Japanese garden design is a *complete* approach to design via an appeal to all the sense organs seems clear. Within the studied gardens presented in this chapter, the strategies employed within both spaces to create particularly memorable and rarefied audio-visual experiences are not circumstances of chance, but expert manipulations of landscape features in the knowledge that the auditory experiences that will result will enable another dimension of the garden to be revealed. As such, particular topographic features of the sites are exploited such as the dramatic drop-off at Kyu Furukawa Teien where the lower valley that contains the Japanese garden sits in a favorable bowl-like condition to allow for sounds from the *ootaki* to travel across the *shinjiike*. Materialities that reflect sounds are favored too, in particular areas such as at the *karetaki* and the *shiraito no taki* at Koishikawa Korakuen where local sounds are reflected and amplified by rocks and hard surfaces so as to create a unique micro environment in which other exterior city sounds become masked. These types of approaches to design in which there is a concentration on providing the visitor with a number of opportunities to both hear and see vistas and features is commonplace.

The natural question that arises then is how this model, as exemplar, and identified as to its structural traits, can lead to other manifestations or the production of other exemplars. If the focus then of this chapter has been through the lens of analytical approaches that attempt to extract pertinent insights into the basic structures that underline Japanese garden design, I present in the following chapter a different tact—that of a series of creative and artistic works that explore how the insights into the structure of Japanese garden design explored so far may provide an intellectual currency for artistic production—that is, a series of methods that attempts to *re-produce* the Japanese garden within other media. But rather than simply removing any definitive research question in favor of pure artistic expression, the projects undertaken instead seek new creative ways in which spatial analysis can be used to drive artistic synthesis. The approach is then one in which the precedent of Cage's usurpation of the garden at Ryōan-ji in his musical work of the same name might be revisited in light of the insights and findings I have presented in this chapter. That an artistic work or artistic process might partake equally in usurping, integrating or finding influence from a range of more formal methodologies for spatial analysis is a goal that I use as a means to examine not so much how the Japanese garden might solely be re-invented within a secondary media or as artistic artifact, but moreover, how the approach of the *niwashi* might be re-applied to a secondary media. How then might new materialities (virtual or ephemeral) affect the act of composition? The act of garden design involves *thinking through* how a library of materialities, forms and textures can be combined in a fashion that produces known audio-visual encounters. These considerations are then embodied in myriad other elements that frame notions about site, scale and the semiotics of landscape forms as viably malleable tools for creating a rarefied garden experience. This represents then what I nominate as a form of *spatial thinking* in which particular structures, combinations and associations of elements within a garden designer's palette becomes a utility for which, under an experienced hand, come together to produce particularly striking landscapes. How then might this framework and approach to spatial thinking find an outlet within other media, artifacts or artistic processes?

After Cage

Transmediating the Japanese Garden

Introduction

Though John Cage can be most easily identified as finding the influence of Japanese garden design an irresistible artistic force for adaptation, other high profile architects such as Frank Lloyd Wright, Mirei Shigemori and Mies van der Rohe have been equally lured. Following in a like manner to Cage, Shigemori similarly examined and translated the Japanese garden through a variety of contemporary temple gardens and sculpture parks that sought to appropriate spatial dynamics via various material manifestations and geometric transformations common to Japanese garden design.[1] For Lloyd Wright, and van der Rohe, the influence lingered through a more subtle aesthetic approach to spatial composition and an acute attention to detail, though remained an endearing and oft identified characteristic of their work.

As I have argued previously, the design of the Japanese garden and its use of natural materials, site topography, controlled variations of texture and number, and expert manipulation of the viewer's perceived presence, produces particularly balanced and subtle multi-sensory encounters. This type of design mastery then has naturally led those aforementioned, and numerous other contemporary artists, to explore the process of translation of the Japanese garden as a means to curate and re-produce similar types of multi-sensory experiences within a range of environments.

Following then in the footsteps of these artists and thinkers, I will present in this chapter my own artistic works that build and perhaps also attempt to tear down Cage's precedent, while at the same time seeking to shift the current focus on purely visual design elements within Japanese gardens as points for creative adaptation. Although recent sensory studies conducted on Japanese gardens have positioned them as generators of unique spatial paradigms, they have mostly focused on the visual information contained within them. For example, the work of Miki Kondo and Ry-

1 Mirei Shigemori, *Kare-sansui* (Kyoto: Kawara Shoten, 1965).

uzo Ono examines the spatial arrangements inherent in Japanese gardens as potential rhythmic structures.[2] This point has also been similarly suggested through the work of Kiyoshi Furukawa, Haruyuki Fuji and Yashuhiro Kiyomizu,[3] though both investigations negate a true multi-sensory examination through a privileging of visually derivable data in their respective examinations. Yosuke Kinoe and Hirohiko Mori[4] have speculated on the connection between Japanese cultural values concerning *soto* (outside) and *uchi* (inside) as generators of spatial paradigms within Japanese gardens, while the work of Gert van Tonder (of which I discussed briefly in chapter 2) has relied on examining the geometry of rock placements at Ryōan-ji in order to point out unseen spatial structures that mimic naturally occurring patterns.[5] Additionally, more traditional investigations into the interpretation of the spatiality of a Japanese garden by Lorraine Kuck[6] and Bernard Berthier[7] have forwarded both meta-physical and aesthetic qualities as important factors in governing a garden's inherent spatial rules.

Given my study of the Japanese garden design thus far has emphasized them as not only a function of ocular articulations of space, but similarly, highly composed aural negotiations of site, the question that has arisen is what are the implications for a *transmediation* of the proclivities of the Japanese garden for sound designers or other composers? Can the spatial aesthetics of Japanese garden design provide a means to drive sound design or musical composition, or become a catalyst for complimentary aural architectures? What is the legacy of Cage's *Ryoanji* precedent and how may it be extended or continued, and furthermore, how might other innovative and unique approaches to the translation the Japanese garden into visual or auditory works arise?

While I have thus far applied various formal methodologies and mathematical abstractions as a means for the analysis of the spatial and auditory nature of the Japanese garden design, I will in this chapter turn to creative and artistic expressions in an effort to uncover and identify those underlying qualities that may arise or consequently become lost in the translation between traditional garden architecture and translated artistic artifact. As such, in this chapter I discuss recent auditory works

2 Ryuzo Ohno and Miki Kondo, "Measurement of the multi-sensory information for describing sequential experience in the environment: an application to the Japanese circuit style garden," in *The Urban Experience: Proceedings of the 13th Conference of the International Association for People-Environment Studies* (Manchester: E & FN SPON, 1994), 425-437.

3 Kiyoshi Furukawa, Haruyuki Fujii and Yashuhiro Kiyomizu, "A comparative study on the relationship between music and space experience, based on tempo at a circuit style garden, *IPSJ SIG Technical Reports* 19 (2006): 7-12.

4 Yosuke Kinoe and Hirohiko Mori, "Mutual Harmony and Temporal Continuity: A Perspective from the Japanese Garden," *ACM SIGCHI Bulletin* 25/1(1993): 10-13.

5 Gert van Tonder, "Recovery of visual structure in illustrated Japanese gardens," *Pattern Recognition Letters* 28 (2007): 728-739.

6 Lorraine Kuck, *The world of the Japanese Garden* (New York, NY: Weatherhill, 1968).

7 Bernard Berthier, *Breaking the cosmic cycle: religion in a Japanese village* (Ithaca, NY: Cornell University, 1975).

(that is, sound installations and compositional theory) as well as sculptural works and synthetic environments (or what I have termed proto-architectures). Together, these projects form an artistic investigation and counter-balancing arm to the previous analytic excursions. But while they are cast within an artistic framework, they at the same time seek to rigorously usurp the spatial predilections of the Japanese garden design through appropriations of some of the formal methodologies introduced in the previous chapter (such as FCA), while introducing new techniques such as Voronoi tessellation. I thus extend the scope of the overall study by introducing project work based on the already analyzed *kare-sansui* at Kyoto's Ryōan-ji temple, the *otowa no taki* from Koishikawa Korakuen, $\mathfrak{B}(\mathbb{K})$ of Kyu Furukawa Teien as well as the small stroll garden *sesshutei* by the famous Muromachi-era monk painter Sesshū Tōyō.

The modus operandi of the project work presented in this chapter then is driven by an intense interest and influence from the theory of *transmediation* by Marjorie Siegel.[8] Here, Siegel defines transmediation as when a sign system in one media is re-appropriated into another. I thus use Siegel's notion of transmediation as a methodological guide, and place Cage's *Ryoanji* precedent as an important exemplar or paradigm. But rather than fixing the boundaries and clearly defining transmediation as only encompassing one particular methodological approach for artistic production, I instead seek to experiment by way of developing myriad approaches. The artistic processes then that arise from this particular framework for working inevitably seek a re-seeing and re-hearing of the Japanese garden. In this manner, transmediation is framed as an ongoing *search*, such that ever-new paths for creative practice may emerge. Under these auspices, the investigation remains open, though securely focused on continually developing innovative ways of uncovering the spatial predilections within the Japanese garden through the act of spatial making and in particular, *spatial thinking.*

Beyond Ryōan-ji

A first foray into the world of transmediating the sign system encapsulated within the Japanese garden into another spatial composition must inevitable pass through, or reference, a known exemplar. The most appropriate case in hand for my investigation is naturally Cage's *Ryoanji*. As a starting point then, the method of Cage's transmediation of the *kare-sansui* into a musical analog (as previously discussed in chapter 2) serves as a methodological paradigm, for which my own re-interpretation and extension within a framework of Cagean aesthetics seeks to provide an initial point of

8 Majorie Siegel, "More than Words: The Generative Power of Transmediation for Learning," *Canadian Journal of Education/Revue Canadienne de l'education* 20/4 (1995): 455-475.

departure, or perhaps nexus for the many other projects and creative investigations I will later discuss in this chapter.

Indeed, Cage's interest in transmediating the *kare-sansui* at Ryōan-ji placed a discrete emphasis on producing a musical equivalent to the elements of focus within the original garden space, and hence the construction of a narrative for understanding how the garden works conceptually, and therefore how it might be 'heard' as music. Rather than developing a compositional approach that derived or sampled information from the source itself, perhaps like those works he based on James Joyce's *Finnegan's Wake* (e.g., *Writing for the second time through Finnegan's Wake*) or the music of early American composers Supply Belcher (*Some of 'The Harmony of Maine'*) and William Billings (*Hymns and Variations*), the process he used in *Ryoanji* was a new appropriation which was closely related to the work in printmaking and etching he had recently undertaken at Crown Point Press in the early 1980s. Indeed, the only direct reference of any environmental sampling from the site of the *kare-sansui* garden occurs with the use of the number fifteen (which totals the number of rocks within the garden). The process of composing *Ryoanji* then was goal-oriented, and representational, unlike many of his previously purely indeterminate works from the 1960s such as *Variations II* (1961). Indeed, as a model of working for abstract means, *Variations II* provides an excellent contrastive example that highlights a method of composition that is non-representational and non-programmatic.

The graphic score of *Variations II* is composed of five transparent sheets with a single point in each, and further sheets that contain lines. Working more as a tool rather than discrete musical score (in fact more a means to make a score), the performer(s) is to overlay the sheets and after assigning specific parameters to the lines (i.e., frequency [pitch], duration, point of occurrence, timbre, structure), drop perpendiculars to the points, and through a system of measurement, obtain data about the sound events. As with most of Cage's works from the period, this operation can be repeated freely to obtain a large enough data set for a performance of any length for any number of people. In analyzing the work, Thomas Delio[9] notes its structure as peculiar and multifarious, and thus suggestive of an infinite number of formal structures at any one point. Indeed the musical work *Ryoanji* shares a heritage with *Variations II*, at least within the method to which the construction of the work involves producing a musical outcome from some non-musical template to which relevant information is obtained. For *Ryoanji*, Cage's personal collection of fifteen rocks provided the outlines for the thematic content, while *Variations II* works in a more disconnected fashion in which the parameters of a musical object are found through a series of measurements.

9 Thomas Delio, "John Cage's *Variations II*: morphology of a Global Structure," *Perspectives of New Music* 19/1-2 (Autumn 1980/Summer 1981): 351-371.

Using measurement of points to lines, defines for the work a design or compositional space that is predicated on the relative metric used for measurement, as well as the method to which the values of measurement are translated into musically meaningful data. In contrast, the method to which *Ryoanji* was realized produces a limited window for variation, and moreover, the unchanging nature of the topology of the template rocks infuses the corresponding transmediation with a perceivable or even auditory trace.

Re-Ryoanji: Towards a New Spatial Model

In an effort then to perhaps both re-appraise and re-interpret Cage's transmediation of the Ryōan-ji garden into a musical work, I have proposed a re-thinking of the piece that seeks to look back into the geometry of the *kare-sansui* and adapt the nature of Cage's realization methodology of *Variations II*. This approach to transmediation positions a subsequent way of usurping the *kare-sansui* at Ryōan-ji through applying a more formalist or eco-structuralist approach to composition. Rather than adopting a similarly representational position such as Cage did in utilizing a symbolic fifteen stones that were traced around to generate melodic elements, the non-representational abstractions of measurement used in *Variations II* are applied to intrinsic environmental data of the site. By usurping Cage's goal of a media translation between Japanese garden design and musical composition, *Re-Ryoanji* is a methodology for a closer connection between the garden and the intended musical output.

The tracing of collected stones by Cage (see Figure 5), was a topological appropriation or inspired process that facilitated an aesthetic or poetic connection between the process and the outcome. Given that the stones were not from the original garden, how might a musical environment be constructed from a usurping of environmental conditions? Gert van Tonder and Michael Lyons's examination of the visual effects of the *kare-sansui* at Ryōan-ji is revealed through a Medial Axis Transformation (MAT) model that highlights the nature of the relationship between rock groups and the raked gravel that lies between them, and their connection to the temple architecture. MAT is commonly used in computer science (e.g. for mesh generation, image analysis and computer vision) to reveal object shape topologies (or topological skeletons) in Euclidian space. Voronoi tessellation is a similar method by which the relationship between points on a plane (Delaunay Nodes), boundary areas of equidistant points between Delaunay Nodes (called Voronoi Edges), and the intersections of Voronoi Edges (called Voronoi Nodes) are visualized through a tessellation diagram.

In using a similar approach to Van Tonder and Lyons, a Voronoi tessellation of the *kare-sansui* at Ryōan-ji reveals areas of influence from each rock grouping and its constituents. Figure 27 shows a schematic diagram of the *kare-sansui* in which each point (or Delaunay node, Dn^i) is representative of one of the fifteen rocks of the garden, which is further comprised of 5 larger groups (g^n). Lines (Voronoi edges, ve^i) demarcate areas of influence whereby the collection of points within each line, lie equidistant to each focus point, Dn^i. As Van Tonder and Lyons have also asserted, there is a drawing in of the focus of the rock groups, that points towards the larger empty area that roughly divides the garden in half.[10] The Voronoi diagram is a useful tool in deconstructing the space in terms of its Euclidian geometry and revealing the abstract nature of how the elements of the garden relate to one another.

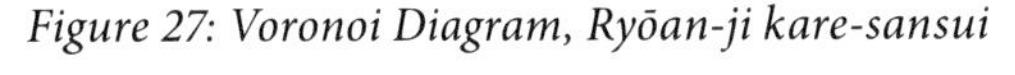

Figure 27: Voronoi Diagram, Ryōan-ji kare-sansui

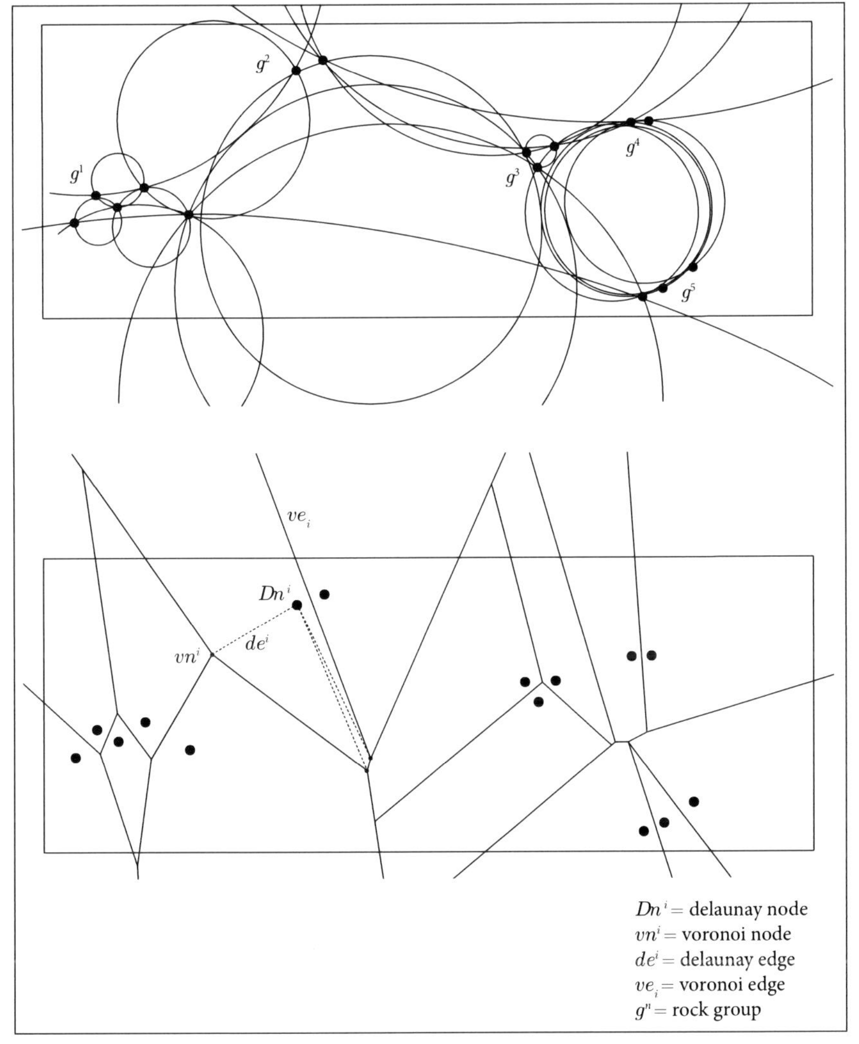

10 van Tonder and Lyons, "Visual Perception in Japanese rock garden design," 357.

The construction of Voronoi edges is a method to which the processes used in *Variations II* may be re-applied within the context of the Voronoi tessellation diagram. For the template created in *Variations II*, lines of the transparencies are assigned to the musical parameters: frequency (pitch), duration, point of occurrence in given point of time, timbre, structure. Qualifying the Voronoi edges with one of the above parameters produces an environment in which measurement of an event (any rock-group focus point, dn^i) to the intersection of hyperplanes (Voronoi nodes, vn^i) produces a value (also known as the Delaunay edge, de^i) that exits as an equivalent datum element to values obtained through the congruent methods of *Variations II* (see Figure 28). Using such an approach enables a usurping of environmental data from the garden site. If, as in *Variations II*, points from the diagram are interpreted as representing sonic events, then the boundaries to which the Voronoi cells are articulated, (the Drichlet domains) may be used to obtain data that defines these events.

Figure 28: Derivations of de^i in g^1

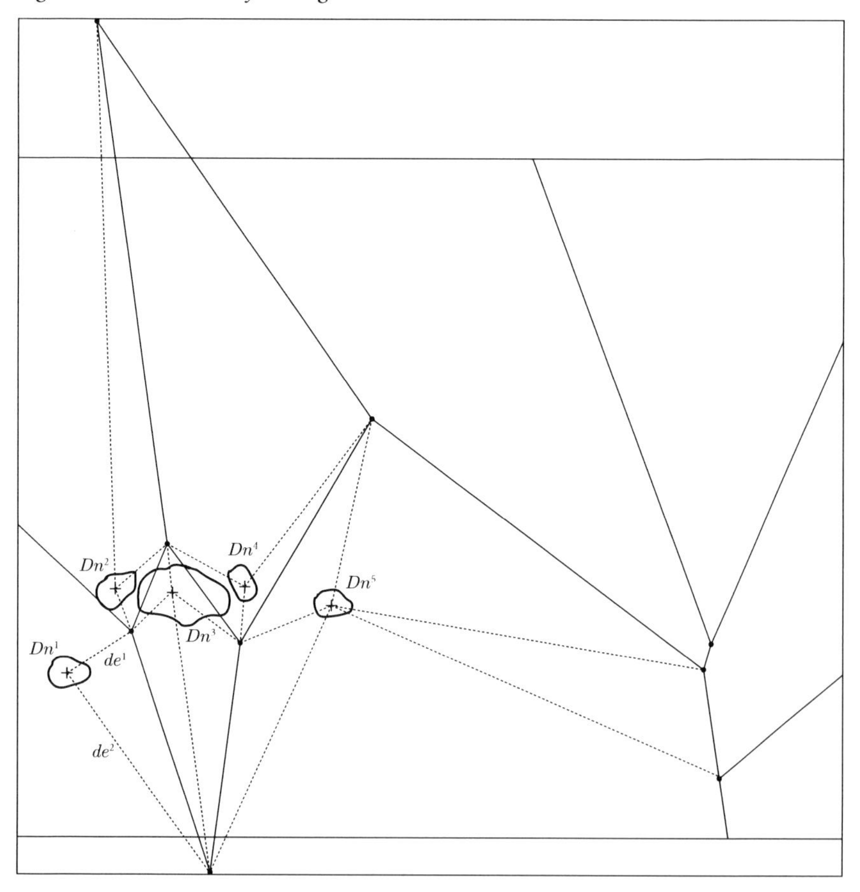

Using the Voronoi nodes as the points to which measurement is actuated enables a more complete spatial-specific context, where the number of values that define Dn^i equals the number of nodes (vn^i) contained within each cell. Using Ryōan-ji's Voronoi nodes (see Table 5) as a template to which musical parameters can be obtained provides a type of eco-structuralist framework in which the garden's attributes are utilized.

Table 5: Delaunay node data set

Group	Delaunay Nodes	vn^i (Voronoi node index)
g^1	Dn^1	$i=\{1, 2\}$
	Dn^2	$i=\{1, 2, 3\}$
	Dn^3	$i=\{1, 2, 3, 4\}$
	Dn^4	$i=\{1, 2, 3, 4\}$
	Dn^5	$i=\{1, 2, 3, 4, 5, 6\}$
g^2	Dn^1	$i=\{1, 2, 3, 4\}$
	Dn^2	$i=\{1, 2\}$
g^3	Dn^1	$i=\{1, 2, 3, 4, 5\}$
	Dn^2	$i=\{1, 2, 3, 4, 5, 6\}$
	Dn^3	$i=\{1, 2, 3\}$
g^4	Dn^1	$i=\{1, 2, 3\}$
	Dn^2	$i=\{1, 2\}$
g^5	Dn^1	$i=\{1, 2, 3, 4\}$
	Dn^2	$i=1$
	Dn^3	$i=1$

Using the Voronoi nodes as the points to which measurement is actuated enables a more complete spatial-specific context, where the number of values that define Dn^i equals the number of nodes (vn^i) contained within each cell. Using Ryōan-ji's Voronoi nodes (see Table 5) as a template to which musical parameters can be obtained provides a type of eco-structuralist framework in which the garden's attributes are utilized.

As an example of continuing the process through to obtain discrete musical information, the diagram of Figure 28 is an available map to which scale and data can be obtained directly from the source, in this case for Dn^1 of g^1:

where g^1 *contains points* $\{Dn^1, \ldots, Dn^5\}$

and $de^1 \in Dn^1 = 10,$

$de^2 \in Dn^1 = 35$

The values of de^1 were obtained using millimeters, though as in *Variations II* this may freely change between node points and groups (i.e. values for de^i may be defined with a changing metric). Values may even be observed qualitatively (as with pianist David Tudor's first realization of *Variations II*)[11] or as an algebraic, or transformational series of relationships. A complete mapping of the garden nodes of g^1 using millimeters yields the data set of Table 6. The frequency of common values throughout the table is a function of the measurement of nodes to points.

Table 6: Values of $de^i \in g^1$

Delaunay Node	Values of de^i
Dn^1	{10, 35}
Dn^2	{9, 12, 73}
Dn^3	{9, 12, 12, 25}
Dn^4	{12, 12, 73, 32}
Dn^5	{12, 25, 32, 60, 96, 102}

Each of the Voronoi nodes is the intersection of lines of common distance between cell interiors. This type of mapping or reading of the tessellation focuses on the how the garden is a series of empty 'holes', (Figure 27 [top]), to which the content of rocks draws and bounds the nature of the perceptual experience, and points towards both the boundary of the garden as well as the sphere of influence of the rocks themselves. As Mitchell Bring and Josse Wayembergh have noted it is this empty space that is "both poignant with meaning and a crucial factor in the design."[12] Interpreting then the values of de^i and translating them into musically meaningful representations is a common problem to the process of realizing *Variations II*. The numerous ways in which data becomes a musical parameter lies at the heart of the nature of this indeterminate structure. Understanding Table 6 as a mapping of pitch requires an interpretation that may significantly alter the identity of the transmediation. Figure 29 presents an approach that translates the data set into midi-note numbers, and therefore note names, such that a musical end is reached relatively simply.

11 James Pritchett, "David Tudor as Composer/Performer in Cage's *Variations II*" *Leonardo Music Journal* 14 (2004): 11-16.

12 Mitchell Bring and Josse Wayembergh, *Japanese Gardens: Design and Meaning* (New York: McGraw-Hill, 1981), 186.

The Nature of Musical Equivalents

As a spatial model that outlines a methodology for transmediation, *Re-Ryoanji* is an attempt to use both a greater knowledge of the original garden's geometry (or environmental conditions), as well as Cage's established, indeterminate compositional process to extract site data for musical means. While a pitch generating structure, centering on a generic interval transformation from a C4 (middle c) node, was developed in Figure 29 for each element of g^1, any of the Voronoi hyperplanes in the model could be assigned a musical parameter from *Variations II* (i.e. duration, timbre, etc.) to obtain other musical information. Similarly, a transformational network could be invoked from the relationship between Dn^i and g^n, thus creating a greater variety of pitch space through transpositions. What is pertinent for this particular spatial model though is the relationship of the extra-musical to the musical object. In Cage's transmediation of Ryōan-ji, an extra-musical program arises because the intent of the work is to create an analog to the *kare-sansui* (even though the compositional process used chance methods). This fact instills in the piece a sense of narrative, and of the music as pointing and embodying a specific extra-musical object—be it the original garden, or the sense of creating an acoustic ecology: recall Cage's observation "each page is a garden of sounds."[13]

Figure 29: Pitch-class generation sequence

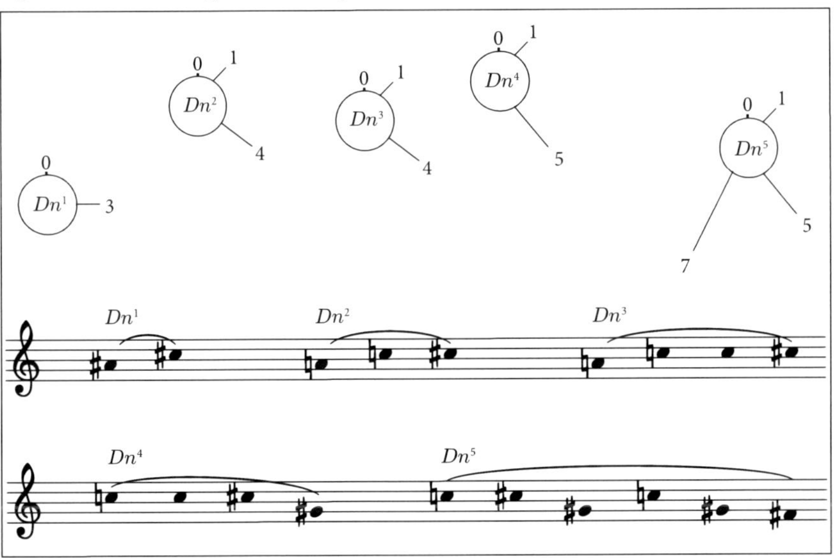

13 Cage, *Ryoanji*, 1.

However, the *kare-sansui* at Ryōan-ji is itself unique and a superb example of the break in Japanese gardening tradition of the time through its aesthetic move away from pure representation to a more erudite embrace of the intangible nature of Zen. If there is meaning in Cage's *Ryoanji*—that the work is a garden of sounds, or points towards how a garden might sound as music—it becomes not unlike yet another process of layering transparencies. The implications of reading the *kare-sansui* in terms of what it is representing tend to turn the garden back onto itself as any famous Zen *koan* (or traditional riddle) might. Therefore the translation produced by Cage essentially layers another 'reading' onto the purely abstract garden, in turn perhaps re-forming how it is viewed, or at least creating a model for how the garden might be read in a musical fashion. Cage's instigation of a musical representation for the space uses all the tools of chance and non-determinedness in an attempt to represent something that is highly composed and completely determined.

Though, as a musical translation of a specific garden, Cage's *Ryoanji* is more poetic than formalist in it's approach. The processes of *Variation II* have been usurped here for *Re-Ryoanji* and framed in what I would argue as a seemingly more appropriate impetus for guiding a transmediation in which the spatial properties of the *kare-sansui* are utilized within a compositional environment. Ultimately then, *Re-Ryoanji* is a model for eco-structuralist approaches to composition and acts as a first pass or initial investigation into how Cage's original *Ryoanji* precedent might be re-thought with a new framework for transmediation. It is then necessarily a product of both *Ryoanji* and *Variations II*. The abstract nature of transmediation, in this case geometry becoming musical pitches, will itself reside as close to the poetic nature of Cage's *Ryoanji* as with any other musical translation of the garden that is based on unheard structure of the site. Perhaps a true musical translation must be understood aurally, and therefore take into account that which the space reveals in its soundscape—the irony of which presents another *koan*-like dilemma by the fact that those sounds native to the site are made by visitors and sources that lie outside of the confines of the garden.

Eco-structural Sound-Space Design

The exploration then of a possible counter transmediation to Cage's approach regarding the *kare-sansui* at Ryōan-ji remains both a point of departure and a valuable utility. As a departure point, the spatial paradigm created by Cage brings to the fore the question of what might a transmediation sound like that utilizes the environmental sound sources of the garden or of its surrounding acoustic ecology? How might the Cage's *Ryoanji*, when viewed as a type of elementary soundscape composition, enable other compositional approaches to emerge that are ecologically based on auditory conditions or auditory site information? Can such eco-compositional strategies for soundscape design in themselves reveal hidden design artifacts or spatial predilections of a Japanese garden?

Indeed for both Barry Truax and fellow soundscape composer Damián Keller, the notion of ecologically based composition is not only a function of the theories that have arisen from acoustic ecology and the perceived utility of the paradigm of soundscape composition, but similarly from the work of James Gibson and Roger Barker in the field of environmental psychology. Gibson's seminal 1950 book, *The Perception of the Visual World* established that the nature of the body's sensory systems is a relationship between perception, stimulation and sensation. Blesser and Salter similarly reference this notion when they suggest that listening is a tripartite encounter in which sensation (detection), perception (recognition), and affect (meaningfulness) provide a basis for which an external environment becomes an internalized auditory experience. But Gibson was not singularly concerned with the phenomenon of listening, but a complete multi-sensory approach to understanding human perception. For Barker, who later continued the work of Gibson, the new field of ecological psychology came to be defined as being concerned with:

"both molecular and molar behavior, and with both the psychological environment (the life space in Kurt Lewin's terms; the world as a particular person perceives and is otherwise affected by it) and with the ecological environment (the objective, pre-perceptual context of behavior; the real-life settings within which people behave)."[14]

The obvious connections then to the field of acoustic ecology are further emphasized in the notion that ecological psychology seeks to examine the relationship we have with the natural, man-built or man-modified environment, and in particular, the impact of both social and physical attributes that such spaces embody—"The physical and interpersonal properties of the environment are distributed in space, and personal environmental space is shaped by the configurations of these properties."[15]

The usurpation then of the theoretical framework of ecological psychology has been the primary impetus, and indeed a rallying point for the techniques of soundscape composition. Unlike the European acousmatic tradition, soundscape compositions seek to specifically reference and leave untouched the original source material for its social, cultural and auditory context. For example, works such as Charles Fox's *wildurban*, utilize sampled environmental sounds that are combined and juxtaposed with synthesized, algorithmically generated events to create a sense of a fractured and new auditory dimension that has obvious allusions to both the natural world and the world of twentieth century electro-acoustic composition. The

14 Roger Barker, *Ecological Psychology* (Stanford: Stanford University Press, 1968), 1.
15 Robert Beck, "Spatial Meaning and the Properties of the Environment," in *Environmental Perception and Behavior*, ed. David Lowenthal (Chicago: University of Chicago, 1967), 18.

combination of recorded sounds (normally from pristine and untouched natural environments) together with sampling processes such as granular synthesis or other digital signal manipulation, give a reliable definition to what has come to be known as the Canadian soundscape school.

But given that the work of Barker and Gibson investigates the relationship between humans and the environment, and moreover the impact of the environment on human behavior, the adaptation of the methods of ecological psychology towards the theory of soundscape composition or ecological sound design has naturally developed around a preferencing for understanding sounding events as isolated flows of information from an environment. But Gibson observes that the physical environment, as an information system, contains data (or *environmental invariants*) that in relation to humans flow through many mediums that are formally defined as any operationally independent segment or channel. Gibson proposes that the human system of perception then, is a multi-sensory one that seeks to adapt and learn to determine the value of the invariants of the medium that conveys that information.

Gibson also argues that as an increase in the frequency of exposure to different mediums develops, so too a greater increase in the ability one has to define the particular sensation that is being experienced. In his essay on McLuhan and Gibson, Lee Frank identifies the nature of the auditory sensation:

"Vibrations through the media of air are picked up by the auditory sensory system. Vibration of a certain range of amplitude and frequency are identifiable as sounds. Media within media. Certain sounds are human speech. Certain human speech is verbal communication. Certain verbal communication is emotional. The auditory sensory system can select any of these media as its information. Its ability to determine the invariants, which define a medium, allows it to select the sensations, which will become perceptions—the sleeping mother hears the baby's cry. A medium, therefore, can contain (abstract from) another by constraining some variable(s)."[16]

In the case then for developing an ecologically based musical work, or in fact another transmediation of a Japanese garden, an important issue not completely explored by Cage in his *Ryoanji* score arises through considering those sounds emanating from the garden itself.

As the human perceptual system is one that can learn and adapt to process the most meaningful invariants of a given medium, obtaining invariants from an environment can be easily re-used within a musical system that re-synthesizes or re-appropriates such data in real-time. In this fashion, the musical system (which is necessarily digital), obtaining information from the physical environment (the analog), becomes its own sensory and perceptual system in which the invariants of the envi-

16 Lee Frank, "Media as input to Perceptual Systems," *www.leefrank.com* 1969, accessed March 18, 2013, http://www.leefrank.com/inquiry/mind/consciousness/media.html.

ronment are re-cast into an ecologically musical realm: in essence an ADC (or analog to digital convertor). This relationship of systems is one in which environmental data, positioned firstly as a musical stimulus, shapes and forms a new auditory experience within the musical sphere—a process perhaps of revealing the 'unheard' within the garden.

In the following discussions I will examine how this over arching notion of ecological sound design has found currency within a series of artistic works. The focus thus becomes how an artistic practice generated through the composition and design of spatial sound environments can be forged through a theoretical and philosophical context, and moreover what further insights might be gained from such an approach. Whereas the focus of the first part of this book has been generally towards the description of a number of methodologies for the analysis of spatial predilections of the Japanese garden, I want to focus somewhat now on the relationship between formal theories and artistic practice. As such, three large-scale immersive sound installations are presented, *Acoustic Intersections*, *Sesshutei as a Spatial Model* and *Shin-Ryōan-ji*. Each work represents a self-contained experimental model that investigates both methods for usurping spatial information as well as how the transmediation of this information itself may reveal new insights into the Japanese garden. In this sense then, the tools of transmediation and the theoretical context from which they are drawn are viewed as not exclusively for artistic exploration, but ones equally capable of generating new knowledge of the underlying structural characteristics of the Japanese garden.

Acoustic Intersections

Initially conceived as an eco-structuralist sound design project for Hakodate Future University's Media Laboratory (Hakodate, Japan) in 2006, *Acoustic Intersections* explores the potential connections and disconnections that arise between the acoustic ecology of a traditional Japanese garden, its environmental sound sources, Schafer and Truax's ideology of soundscape composition, and a desire to explore Blesser and Salter's notion of spatial auditory awareness. The installation also seeks to map out the relationship between the urban landscape and acoustic experience at the *otowa no taki* at Koishikawa Korakuen, and how this particular encounter at the garden might become the impetus for exploring and revealing the underlying acoustic layers at work through utilizing electronic sound projection together with juxtapositions of architectonic volume and scale.

Damien Keller, Ariadna Capasso and Scott Wilson have previously forwarded the notion of eco-structuralist sound design through their sound installation *The Urban Corridor*. Here, Keller, Capasso and Wilson present a multi-modal, urban intervention that focuses on "the relationship between the public, visual and sonic stimuli, and space as a way of bringing to life the effect of the urban landscape."[17] In an approach that the authors describe as ecological, short grains (that is, very small snippets) of sound are extracted from source recordings of the site, such that these grains "provide the basic spectral and micro-temporal features of the sounds to be synthesized."[18] This type of ecological structuring by Keller and colleagues provided a type of model for my own usurpation of the soundscape conditions extent within the *otowa no taki* at Koishikawa Korakuen. But while granular synthesis was the primary approach utilized in *Urban Corridor* my own development of *Acoustic Intersections* moved towards a real-time analysis and resynthesis method. Keller and colleagues opted for using granular synthesis, a technique in which very small samples of sound (ca. 1-50 milliseconds) are layered and combined, often playing at different speeds to achieve myriad possibilities for timbre, pitch and rhythmic inflections. In developing a different approach to creating an ecologically-based musical system, *Acoustic Intersections* was designed as a real-time analysis tool to which long (20+ minutes) samples of recorded material from the *otowa no taki* could be analyzed for musically meaningful information (that is, frequency, timbre etc.), then from this simple musical invariant extraction method, an equally simplistic re-synthesis and convolution routine, for which this sound layer is then further juxtaposed against the original site recordings.

As I have previously discussed, the site from which *Acoustic Intersections* is based, the *otowa no taki* at Koishikawa Korakuen, is located at the western corner of the garden, for which the notion between that of separation (*soto*), and the counter force of inclusion (*uchi*), pervades the soundscape. Resting within a natural depression, a small waterway literally trickles past a number of stepping-stones (*sawatari*) that lead a path across the northern most tip of a shallow lake that is littered with smooth stones (Figure 12). From the *sawatari* (Figure 16), which deliberately slows the traverse across the waterway, the city is just visible by a series of tall buildings that border the western stonewall, but overwhelmingly, there is a sense of the aural chaos of the city as it periodically invades the confines of the area.

Here, and as a result of the modernization of Tokyo, an experience that could not of been afforded by the original owners of the garden occurs. Traffic and construction noise filter through the large trees planted at the perimeter walls, with occasional horns and exhaust noises punctuating the calm and steady surroundings of the *otowa no taki*. The soundscape at this point in the garden is in constant flux from active acoustic agents operating as *uchi* objects of inclusion and familiarity (the sounds of garden visitors and the birds that perch on the low stones of the lake) together with those *soto* sounds (from the busy city streets). Because of this flux of agents manifes-

---

17 Damián Keller, Ariadna Capasso and Scott Wilson, "Accumulation and Interaction in the Urban Landscape: 'The Urban Corridor,'" in *International Computer Music Conference Proceedings* (Göthenburg: ICMC, 2002), 295.

18 Keller, Capasso and Wilson, "Accumlation and Interaction in the Urban Landscape," 296.

ting the acoustic space, the idyllic and highly composed visual surrounds are both a point of extroverted drama and inwards contemplation.

For this reason, the *otowa no taki*, was chosen to be the focal point for a sound installation that bases it musical structure and form on the environmental invariants (carried in the medium of sound waves) that pervade the immediate area, and the area that lies just beyond the garden's western boundary. As a site for analysis, the *otowa no taki* is a complex auditory experience: while subtle in its textures and ambient sound levels, it is an auditory example of both a man-built and natural environment—i.e. it lies both as a *soto* extension to the city, and as an *uchi* relation to the greater garden.

Software Design

The intent of *Acoustic Intersections* then is to gather as much fundamental information as possible via the medium of sound from the *otowa no taki* at Koishikawa Korakuen (that is, both the perceived or *heard*, and embedded or *unheard*) as a way of better understanding the acoustic dimensions of the site. The environmental acoustic data, or invariants, from the site are positioned as a template for which a transmediation of the template into a spatial sound installation is actuated through a translation of those invariants into fundamental musical parameters such as frequency, amplitude, and duration. Using techniques of acoustic analysis, spectral data extraction, re-synthesis and convolution, the resulting sound installation acts as a musical transmediation of the soundscape of the *otowa no taki* into an augmented soundscape environment.

The real-time graphical programming environment Pure Data (Pd) was utilized as a means to analyze, extract, and then re-synthesize the environmental invariants of the *otowa no taki* due to the software's employment of a number of native objects and programming routines that quickly enable eco-structuralist approaches to sound design and composition to emerge. In particular, Miller Puckette, Theodore Apel and David Zicarelli developed both the `bonk~` and `fiddle~` objects for real-time analysis of audio signals in Pd. The `fiddle~` object operates as both a mono or polyphonic maximum likelihood pitch detector and has the ability to output: a floating point pitch value which is output when a new stable note is found, a routable message which is output conditionally on attacks whether or not a pitch is found, between 0 to 3 lists (each giving the pitch and loudness of a pitch track), a continuous signal power in dB, and a another list which iteratively sends triples marking each peak's index frequency and amplitude.[19] This functionality means that those most musically relevant pieces of information can be extracted from any sampled site recordings from the *otowa no taki* as a real-time data stream. But more interestingly, the information that is collected is not filtered by the limits of the human auditory

19 Miller Puckette, Theodore Apel and David Zicarelli, "Real-time Audio Analysis Tools for Pd and MaxMSP," in *International Computer Music Conference Proceedings* (Ann Arbor: ICMC, 1998), 110-111.

system—which can only detect frequencies between ca. 20Hz-20KHz. In fact the information that is collected from the site becomes *absolute* and not a function of our ability to find meaning or understand the medium. As Gibson points out: "For man there are classes, sub-classes and instances of identifiable sounds in countless variety, even if one limits consideration to those that can be named."[20]

Figure 30: Installation sketch and channel mapping, Acoustic Intersections

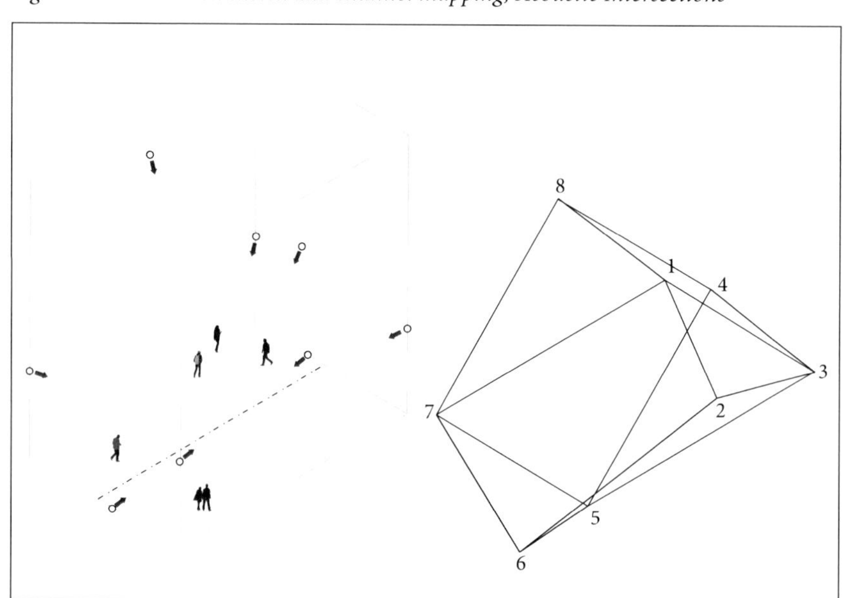

The sound installation itself (Figure 30) was firstly presented at the Media Lab at Hakodate Future University, Hakodate Japan in 2006. *Acoustic Intersections* is presented purely as an auditory record of the *otowa no taki*, though stratified through three juxtaposed yet interacting acoustic spaces. A raw (that is, unedited) twenty-minute audio recording from the *otowa no taki* was captured via a spatial seven-channel recording apparatus that was then played back into the Media Lab through loudspeakers situated in a geometry that essentially mirrored the original microphone positions. This multi-purpose room has a particularly high ceiling (10m), though relatively small footprint (20m x 10m). Juxtaposed over this original site recording is a four-channel composed soundscape layer that transforms (in real-time) the raw recording being played back, while at the same time adding on a radical convolution effect (Cathedral reverberation). This second layer operates via a Pure Data patch in which the raw recording from the *otowa no taki* is analyzed regarding information such as pitch

20 James J. Gibson, *The Senses Considered as Perceptual Systems* (Boston: Houghton Mifflin,1966), 89.

(frequency), timbre (spectrum) and rhythmic information (sound file onsets) coming from the sounds of the site and its surrounds (which of course includes sirens, cars sounds, etc.). Given much of this information lays outside of the 20Hz-20KHz hearing band, various transpositions and timbrel adjustments are made so that the resulting pitches appear within an audibly perceptible range. Essentially, and in the same manner as the presence of the birds and their sounds at the *otowa no taki* creates an acoustic encounter through their activation of the site, this four-channel layer literally 'feeds' from the raw site recordings. The resulting transformation of the raw site recordings into the composed soundscape layer can best be described as a slow moving series of interacting melodies.

The intended effect then of the installation comes from the juxtaposition of acoustic environments, which is an iteration of the nature of the original site where there is a nexus between the *natural* and the *man-modified*. The small four-channel composed soundscape layer is embedded in an *uchi* relation to the larger seven-channel one which itself is embedded within the confines of the Media Lab. But the name of the installation, *Acoustic Intersections* also points towards the fact that the three acoustic spaces within the work are not only distinct, but similarly of the same realm. The sounds produced by the work, as it were, are both *in* space and a *product* of the space of the Media Lab. This is because the acoustic signature of the original recording, an outdoor space, is being played back within an interior space—though the loudspeakers are located within the Media Lab, the sound sources of the recordings are perceived as being located well outside of the confines of the space itself (perhaps even kilometers away—the sounds of the city). To aid this simulation, the microphone positions of the original recording are reflected in the geometric setup of the loudspeakers within the space, as well as an audio rendering technique called B-format Ambisonics which provides a full 3-D soundfield experience. But as Blesser and Salter would also contend, the container of the Media Lab acts in a passive fashion via its volume and materiality, thus prescribing a distinctive behavior on the sound projection and in turn the listening experience. In this sense the room is used like an instrument in the same manner that the *otowa no taki* is an unsuspecting instrument to the birds.

But the merging of these three acoustic environments is also used as a means to track how the local environmental invariants that lie beyond human auditory perception are auralized, and then aurally tracked against the 'real' sounds of the site. In re-casting such information streams, embedded structures, rates of change and dynamic fluctuations in the original site recording form the basis for fundamental coherent musical structures to emerge in the four-channel soundscape layer. Figure 31 is a sonogram or visual representation of a small section of the audio (mono mix-down) from *Acoustic Intersections* and maps time in seconds (x-axis) versus frequency in Hertz (y-axis). The more prevalent (in terms of dB or loudness) a particular frequency spectrum is, the darker the marking. The upper mapping shown is only of the city sounds from the seven-channel site recording. What can be clearly identified here is a rhythmic, narrowly banded sound signal that gradually appears (ca. 900–1000 Hz). This is a siren from a nearby passing car. The lower mapping in Figure 31 combines the upper sonogram of the seven-channel site recording with the four-channel composed soundscape layer. In the sonogram, it can be clearly seen how the soundscape layer accentuates particular frequency spectrum bands from the seven-channel recording, thus providing a rhythmic and melodic counterpart to the diffuse and wide-ranging nature of the city sounds layer. Hence the nature of the site recordings informs the resulting musical aesthetic: a direct eco-structuralist process of environmental information transformation. The composed soundscape layer created from the Pd patch acquires the original soundscape's *aural dimensions*.

Figure 31: Sonogram, Acoustic intersections

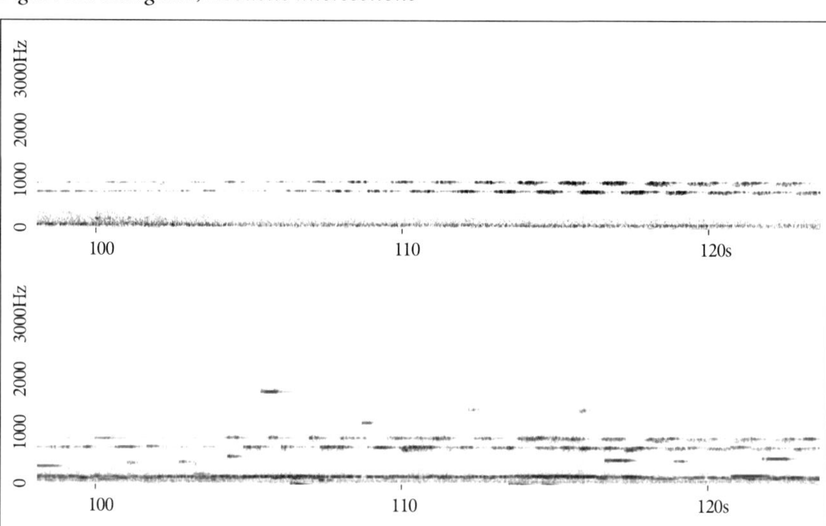

Spatial Meaning

Because of the embedding of the composed soundscape layer within the seven-channel raw source playback, obvious sound events that appear from the sampled garden space are musically echoed within a 'virtual' (that is, musically transformed or transmediated) space (e.g., the presence of passing sirens and car horns etc.). Frequently, though, the transmediated garden's acoustic invariants that appear within the 4-channel composed soundscape layer have no audible relation to the juxtaposed site recording, producing a situation in which complex rhythmic and thematic material emerges from the four-channel composed soundscape layer with no audible analog from the garden source. This revealing of the unheard from the original garden approaches Robert Becks' notion of "immanent" space, or the notion that space may exist that is not readily accessible through our immediate perceptual facilities.[21] Added to this is the fact that the compositional soundscape layer is programmed to imitate the type of listening experience (or reverberation model) one would expect to hear within a Gothic Cathedral. *Acoustic Intersections* then is one that poses the question of what is inside and what is outside, how may an interior architectural space become a passive instrument, what are the dimensions of the acoustic spaces being produced, and within which acoustic space are we as listeners located?

But what is also explored within *Acoustic Intersections*, as a transmediation of the auditory qualities of the *otowa no taki*, is that which R. Burton Linton defines as the relationship between *unity*, *vividness* and *variety* as aesthetic descriptors of landscape experience. Linton further argues on the importance of recognizing the interlacing of *unity*, *vividness* and *variety* and how this connectedness plays an important role in human appreciation of the environment:

"*unity* is the quality of wholeness in which all parts cohere, not merely as an assembly but as a single harmonious unit [. . .] *Variety* in its simple form, can be defined as an index to how many different objects and relationships are found in a landscape [. . .] while *vividness* is that quality in the landscape which gives distinction and makes it visually striking."[22]

Adapting these definitions in regards to the soundscape of the *otowa no taki* at Koishikawa Korakuen provides a method to which the *variety* of sound objects (located both *uchi* and *soto*) are transmediated and therefore can be heard anew as usurped musical materials structured due to the rhythmic *vitality* and a *vividness* (positioned now in terms of auditory qualities) of on-site textual fluctuations. This notion then of 'crafting' an eco-structuralist sound design (by whatever means), as opposed to strictly adapting or translating an extant environmental one, naturally highlights the function and intention of *Acoustic Intersections*. As Schafer and Truax have forwarded previously, soundscape composition is not only a means for artistic ends, but also a useful tool for revealing inherent ecological conditions that connect an auditor to an acoustic ecology. The production of *Acoustic Intersections* then has been a prelude and exploratory investigation into the soundscape of Koishikawa Korakuen and the design implications of developing a compositional method that instigates a macro-level process of eco-structuralist sound design.

The primary interest of the structural design of *Acoustic Intersections* was to create a paradigm to which environmental invariants carried through the medium of sound could be transmediated into a musical context, for which the resulting sonic outcomes heard against the sources from which they were formed. As a counter to

---

21 Beck, "Spatial Meaning and the Properties of the Environment," 18.

22 R. B. Litton, "Aesthetic Dimensions of a Landscape," in *Natural Environments: Studies in Theoretical and Applied Analysis* (Baltimore: John Hopkins University Press, 1972), 284-86.

Cage's *Ryoanji* precedent, this approach acts as a model of Schafer's use of soundscape composition. That is, the configuration of *Acoustic Intersections* approaches Truax's contention that soundscape composition be a platform for enabling what Blesser and Salter have called a "spatial auditory awareness" about a source acoustic environment such that the listener is ecologically placed or at least cognizant of the spatial environment that is being referenced. But Truax and Schafer also frame soundscape compositions in terms of their viability for revealing insights about an environment that other more traditional modes of analysis may not. In the case of *Acoustic Intersections*, such insights have arisen through the concepts of *soto* and *uchi* in relation to both the immediate acoustic surroundings of the *otowa no taki* and its relationship to the larger soundscape of which it is embedded—that of the city of Tokyo. The notion that the intended sound sources of the bathing birds, presented as *soto* interlopers in the garden, assists in marking an acoustic territory of the site is also offset with those other modern, artificial sounds of the city that today find an equal presence. That these sounds combine in such a fashion points to the fact that the acoustic horizon of the site is in continual flux and that the modern-day listening experience of the garden is one that must necessarily accommodate a host of unintended sources. But what *Acoustic Intersections* reveals of this relationship is that the modern-day auditory experience at the *otowa no taki* is an accentuation of the original acoustic dimensions of the site. Whereas the material and topographic qualities of the site would of created an original auditory experience in which the notion of *soto* and *uchi* may have been embodied within the relationship between the fixed physical qualities of the site (*uchi*) and the visiting birds (*soto*) 'bringing in' sounds, modern-day Tokyo has produced an acoustic signature that has itself crept into the garden. *Acoustic Intersections* can then be cast as a tool that reveals the complexity of the spatial and auditory juxtapositions at the *otowa no taki*, and how the wealth of auditory information that constitutes, shapes and guides the behavior of those sounds and their movements *produces* space.

But as a portable programming construct, *Acoustic Intersections* can also be framed as a transmediating machine that may become applied to myriad other acoustic environments. As a comparative analysis tool then, the unchanging nature of such a musical transmediation machine means that the assessment of *any* acoustic environment can be delivered purely through the tools of analysis within the context of a soundscape composition. Spatiality, texture, rhythm and intensity of the source environment against the transmediated musical environment may be qualitatively judged. *Acoustic Intersections* then represents a counter approach to that of Cage's *Ryoanji* transmediation in that it has identified the limitations of neglecting the extant sounding environment of a Japanese garden. The solution though to this alternative approach of usurping or sampling environmental acoustic invariants from a Japanese garden remains a point for further investigation given that the insights that have arisen from *Acoustic Intersections* have bought to the fore the question of *scale* in a Japanese garden. At its most obvious manifestation, the relationship between human scale and the proportions of the human body in relation to the environment provide a measure by which Japanese garden design becomes a subtle art of manipulation and juxtaposition. As is the case in both Kyu Furukawa Teien and Koishikawa Korakuen, the situation of buildings that establish an architectural scale at each garden's entrance is often quickly juxtaposed within a short walk along a path for which those landscape features such as the *otowa no taki* or the *karetaki* are presented as landscapes that have been radically reduced in scale. Added to this are the many bridges such as the *Tsutenkyo* (half-moon bridge) and *Engetsukyo* (full-moon bridge) at Koishikawa Korakuen in which a non-linear morphing of their geometries emphasizes their bowed vertical axis, and thus require slightly more effort to cross them—perhaps this is also used a means to convey the notion of a mythical or intellectual journey. The suggestive *sawatari* that are often located within narrowing paths or for crossing literal or imagined thresholds are also used in ways that suggest that the scale of the surroundings have morphed through diminution or augmentation. Such reductions or expansions though are localized and embedded within the garden in much the same way that the water features are used to create acoustic arenas that articulate a particular part of the garden. Moving through the garden then is a witnessing of the transformation and playful juxtaposition of scale in which features such as hillocks and other topographic manipulations become a series of frames that are constantly shifting in their focus. Scale is a feature that is only established locally within the garden such that two particular areas of the garden may compete in their relative sizes or relative depictions of scale in relation to the viewer. This is further emphasized with the visual technique of *shakkei* as it is masterfully exploited at Adachi Teien. Here the distant mountains that are captured alive and bought into the garden exist at the larger geographic scale and while transcending even the scale of the landscape presented in the garden, they are assimilated in a rather seamless juxtaposition. Perhaps then like Akisato's depiction of a group of visitor's to Ryōan-ji in his 1799 print *Miyako rinsen meishō zue* (Figure 10), the Japanese garden and its juxtaposition of scales plays on the visitor's expectation of *what* is being scaled, the landscape or the viewer? Traversing the garden and encountering the various scenes, each of which may emphasize and utilize various scales through landscape formations, garden ornaments or rock features, only seems to emphasis the notion of the journey as one that is *an unfolding in time*. But with this notion and the relevance of time to the experience of the garden comes the question of how the auditory experience (as a function of time) emphasizes or complements these visual manipulations of scale, size, shape and form, and how this structure or relationship may be transmediated or further usurped. The question then of to what degree does the manipulation of scale have for both the acoustic and visual space of a Japanese garden is a theme that I will be exploring next through the artistic lens of a spatial sound installation based on an early Muromachi garden, Sesshutei by the famous monk-painter Sesshū Tōyō. The installation, *Sesshutei as a spatial model* is presented here as the third iteration and alternate transmediation project to Cage's *Ryoanji* precedent, though purposefully partakes in both the scope, context and experimental approach of *Re-Ryoanji* and *Acoustic Intersections*.

Sesshutei as a Spatial Model

In late 2008 while an artist in residence at the Akiyoshidai International Art Village in Yamaguchi-ken, Japan, I initiated a project to investigate the use of scale in the famous rock garden at the temple Joei-ji in Yamaguchi city. The investigation's end goal was to collaborate with the Yokohama-based innovative loudspeaker manufacturer Taguchi Co. on an eight-channel spatial sound installation that incorporated a series of spatial mappings of the garden at Joei-ji—in essence, a transmediation of the techniques of scale manipulation and landscape articulation that are used in the garden, only now with the intent to understand these approaches within the context of a spatial sound design. Originally conceived as a standalone sound installation, the work, which came to be called *Sesshutei as a spatial model*, developed into a multi-channel performance environment during the course of the collaboration with Taguchi Co. At its end point, the installation became a real-time performance environment in which sampled site recordings from the garden were firstly transformed and manipulated then spatialized through myriad trajectories in real-time according to a framework that utilized the underlying principals of articulation inherent within the source garden. The impetus for *Sesshutei as a spatial model* comes from the northern pond garden located at the temple of Joei-ji (Figure 32 and 33). The garden dates from the Muromachi, and is attributed to the well-known Rinzai monk-painter Sesshū Tōyō (1420-1506). Sesshutei, like the *kare-sansui* at *Ryōan-ji* is viewed most directly from the low veranda of the Buddhist temple. But unlike Ryōan-ji, the garden at Joei-ji utilizes the Muromachi prototypical approach of *chisen kansho teien* in which the earlier Heian era pond garden designed to be enjoyed by boat has been transformed so that the garden is to be viewed from a fixed vantage point. At Sesshutei there is the main viewing point from the temple veranda as well as two further sites on the western and eastern edges of the garden where viewing pavilions were once located. Thus unlike its contemporary Ryōan-ji in Kyoto, Sesshutei combines elements of both Heian and Muromachi prototypes: namely the large central pond, *shinjiike*, which is named (like the pond at Kyu Furukawa Teien) after its shape which is formed in likeness to the Chinese ideogram for heart, as well as a collection of feature rocks that sit in the foreground to the pond. From a seated position on the temple's veranda, both the pond and feature rocks are visible within the immediate undulating plane of the foreground. As Alison Main and Newell Platten note:

"Sesshū built up forms in his paintings with short, angularly related straight lines. He echoed this vocabulary here [at Sesshutei] through the selection of flat-topped, straight-sided rocks: scenes of mountains, ravines, islands and seas are realized as three-dimensional versions of this own paintings."[23]

Sesshū's contemporaries would have readily identified many of these rocks as emulating

23 Alison Main and Newell Platten, *The Lure of the Japanese Garden* (Kent Town (South Australia): Wakefield Press, 2002),165

actual mountains, with the most recognizable being Mt. Fuji. Behind this rather didactic foreground, lotus flowers blossom in the spring, covering the surface of the expansive pond and forming a colorful foil to which the distant Chinese peak Mt. Horai is depicted in its upper right corner. Further behind the mountain is a three-meter high dry waterfall (or *karetaki*) that symbolically feeds the lotus pond (Figure 33) from a hidden (and perhaps heavenly) source.

Figure 32: Sesshutei, Joei-ji, Yamaguchi-shi

What is so striking about the garden is the obvious embedding of scales and the juxtaposition of elements and materials. Sesshū was firstly a scroll painter who studied with Tenshō Shūbun (1414-1563) in Kyoto. He later traveled extensively throughout Japan and China, only designing and making gardens as an outgrowth of his mastery of contemporary Chinese painting techniques (including both the Southern and Northern Song styles). His is particularly noted for his telling angular brush strokes that Thomas Hoover describes as "intricate, dense designs of angular planes framed in powerful lines,"[24] with one of his most famous works, *Winter landscape* (Figure 34), a type of homage to Song Dynasty artist Xia Gui (1195-1224). Like the Sesshutei garden, *Winter landscape* relies on a play of ambiguity of scale by relegating the human presence to one placed on the periphery, and not of central occupation. As Joan

24 Hoover, *Zen Culture*, 126.

Stanley-Barker notes, the early and late paintings of Sesshū celebrated "spatial ambiguity and inconsistencies of scale which had characterized the poetic Shūbun style."[25]

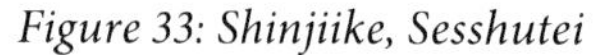

Figure 33: Shinjiike, Sesshutei

Within the monochrome ink painting of *Winter landscape*, a small figure can be made out traversing a steep mountain path in an analogous way to which the obscured garden paths that circumnavigate Sesshutei never overwhelm the primary viewing space from the veranda. But the presence of the human figure in *Winter landscape* also provides a sense of what the immediate scale of the surroundings should be, and thus points to how Sesshū playfully embeds and juxtaposes vastly different scales within the picture space as a means to evoke a sense of otherness. Like the depicted mountain peaks that occur within the foreground view at Sesshutei, within *winter landscape* both the steep mountain pass and two ancient trees frame the immediate surrounds of the anonymous walker. A dramatic shift in scale, not unlike the use of forced perspective at the *otowa no taki* at Koishikawa Korakuen, occurs behind the human figure. Here, the mountain's sharp angular brushstrokes create a sense of vast distance, to which, surprisingly, at its peak, a few trees (including a *literati* styled pine) are painted at the same scale as those occurring at the base of the mountain. Another striking element of the picture includes a detail of a granite cliff, seemingly and somewhat abstractly located behind the mountain. It rises up, disappearing through a soft edge in which the sky and cliff merge. Paul Varley has described this unusual feature of the scroll as a particularly radical departure from the work of Sesshū's teacher Shūbun, while also noting Sesshū's preference for flattening the picture space in *Winter landscape* that:

25 Stanley-Barker, *Japanese Art*, 135.

Figure 34: Winter landscape (1486), Sesshū Tōyō

"Although the scene leads to mountains in the distant background, there is no sense of great depth; and the mountains themselves are not even three-dimensional, but resemble flat cutouts propped against the back of the picture. The most startling part of the winter landscape, however, is in its top center, where a jagged black line appears like a tear in the picture and, next to it, there is an abstract mosaic of surfaces that looks startlingly like the work of a modern cubist painter."[26]

Sesshū completes the landscape with a series of houses and temple, again scaled to the size of the human figure, yet sitting at the rear, and presumably, at some distance from the mountain. Behind the dwellings, the faint outlines of mountain topography more common to China than Japan appear as a suggestive backdrop to the entire scene.

At the Sesshutei garden a similar means for the flattening of the perceived picture space, and the embedding of scale occurs. Numerous *o-karikomi* are suggestive of clouds that form at the base of the miniaturized mountains. These *o-karikomi* and the mountain peaks they congregate around are dissected by *sawatari* that start at human scale, and quickly become much smaller. They provide a path into the mythical landscape that seems to magically shrink the viewer. Behind the foreground mountain peaks, the pond offers yet another group of juxtaposed elements relying on altogether different scales. Here, a number of islands, including one representing a seemingly large boat are set within *shinjiike*. A small wooden bridge leads from the foreground mountain scene to the first island. Further behind the pond lies both the high drop of the *karetaki*, Mt. Horai (at its base) and a number of distant rock formations providing the last feature before the natural barrier of the forest. In a similar manner in which *Winter landscape* is constructed through a process of subtle juxtaposition, each of the elements in Sesshutei seems independent in terms of its scale, yet through the commonalities of color and texture, a striking balance is achieved.

Methodologies for Transmediation

That Sesshū had such a firm yet extremely subtle command of the effects of scale on landscaping in the garden and how such manipulations produce particular landscape experiences in the viewer is a characteristic that may provide a means for which a transmediation into the auditory realm can be codified. But to firstly investigate the manipulation of scale in Sesshutei in a more formal manner may enable a framework or template to emerge that can then be readily used within an auditory context. It is clear from the serene viewing space of the veranda of Joei-ji that the layout of the garden of Sesshutei unfolds in a simplistic yet highly controlled manner. There is a strong sense in the garden of a natural partitioning, yet the dividing up of the space beyond

26 Varley, *Japanese Culture*, 132.

the veranda is subtle and discrete enough because of the use of integrated landscape features such as the rock garden, *shinjiike*, *karetaki* and *o-karikomi*. Indeed, this sense of partitioning within the garden is facilitated through the unfolding of the picture space within three sections beyond the veranda itself. As one would expect these encompass: the immediate point of seating, the foreground, the middleground and the background (Figure 35).

Figure 35: Voronoi development, Sesshutei

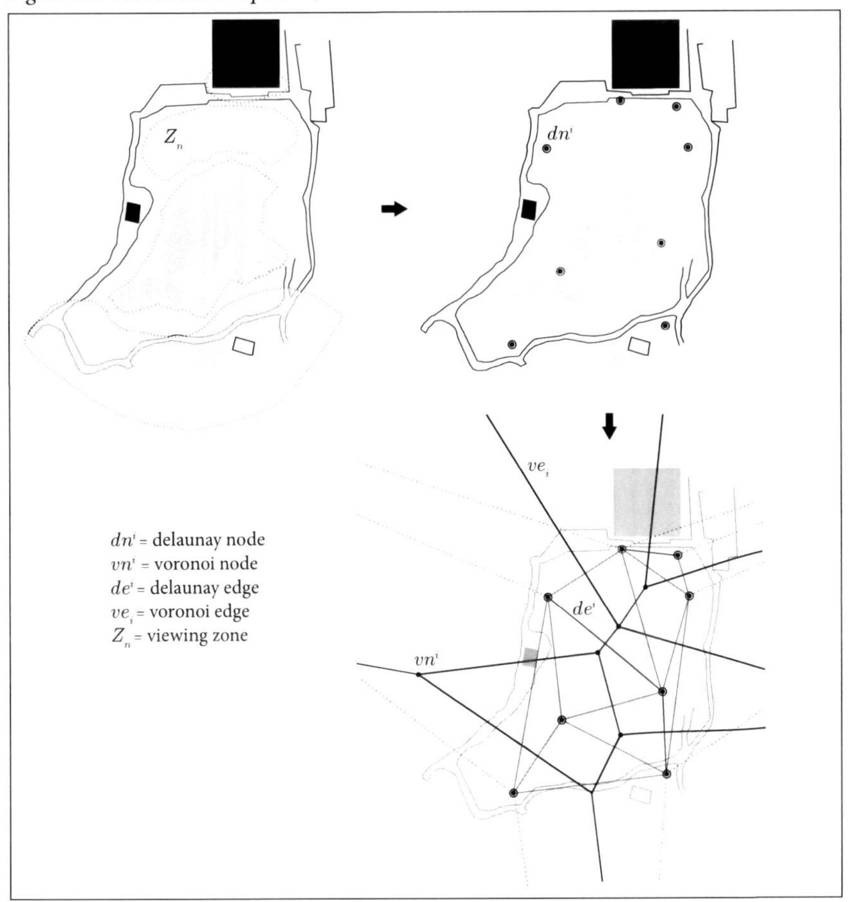

From the veranda of the temple, the viewing space is broken into the immediate surrounds of the viewer at human scale (that is, the veranda and interior of the temple), then the foreground rock formations of famous miniaturized mountain peaks (which include Mt. Fuji), followed by *shinjiike* and its islands, then the *karetaki* and finally the heavy tree cover at the northern end that provides a distinct boundary. Though this naturally occurring partition flows seamlessly from the veranda to the forest beyond *shinjiike*, within each zone Sesshū utilizes unusual juxtapositions of objects and their relative sizes in order to bring a dynamic sense of movement and momentum to the picture space.

Figure 36: Spatial schema of auditory zones

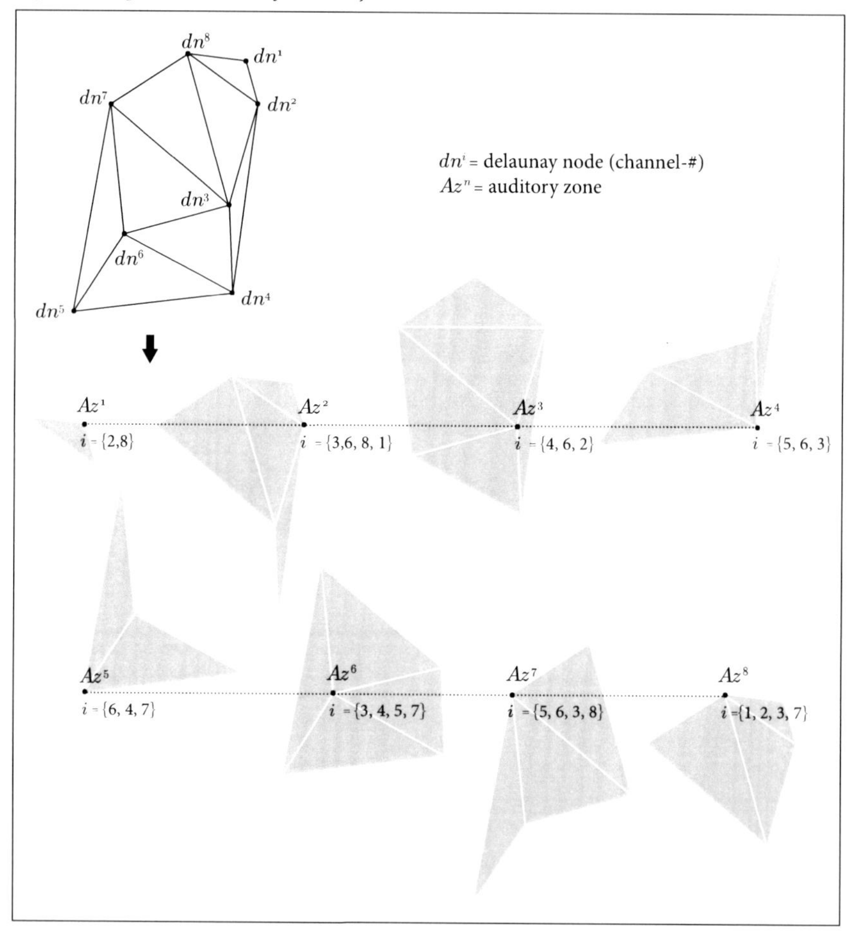

As a way to further investigate this partitioning of the visual space, and, as a means to consider how a strategy for transmediating such a structure into auditory space might occur, the methodology of Voronoi tessellation was revisited. Building on its usurpation in the project *Re-Ryoanji*, the technique was again used at Sesshutei, though instead of utilizing the rocks of the garden as the given Delaunay nodes, a more generalized approach was used. As shown in Figure 35 points (dn^i) were placed within each of the four zones (Z_n) that cover roughly the horizontal range of the viewing space when observed from the veranda.

This mapping approach produces a Voronoi diagram that tracks the spatial relationships between the nominated interior garden points (Delaunay nodes: dn^i) as a function of the delineation of the garden into four zones of the picture space. By assigning each of the Delaunay nodes to corresponding representative points on a plane, the garden can be defined as a metric space to which a Cartesian decomposition may provide a means for the transmediation of the structure of Sesshutei into sound design strategies. As explored in *Re-Ryoanji*, a Voronoi tessellation is a process that accounts for sets of points that lie equidistant to two Delaunay Nodes such that Voronoi edges (ve_i) delineate sub-areas of the plane in question. These sub-areas are also known as Drichlet domains. A Delaunay triangulation, and Boris Delaunay's definition of "emptiness" occurs when the circumcircle of a triangle formed by three points contains no vertices other than the three which define it.

What becomes apparent from the Cartesian decomposition of Sesshutei is the sense that the Drichlet domains that are defined by the Voronoi edges and Voronoi nodes can be considered as not only areas in which points remain equidistant to Delaunay nodes, but areas that may define acoustic qualities or even acoustic arenas—that is, areas for which sound sources are contained or emerge from. This is a naturally occurring characteristic of the myriad micro environments for which the zones are embedded within. The pond for instance contains noises from *koi* and turtles, while the temple itself is a source of Buddhist ritualistic sounds of chanting and instrumental sounds from *daiko* (drums), *rin* (metal bowls) and *mokyugo* (wood blocks). At the far boundary of heavy tree cover come the sounds of the small creek that feeds *shinjiike*.

This method then instigates a means to not only transmediate the site into auditory zones, but also as loudspeaker channel groupings for spatialization strategies (Figure 36). What arises here then is an artistic opportunity to utilize the auditory zones as areas from which sounds might emanate, circulate through myriad trajectories (via loudspeakers) and then even travel through to other zones. Given the sense that each of the Drichlet domains of the analyzed garden are defined by or partition themselves through variously different manipulations of scale, the idea that sounds might also emerge at one scale, then travel to other zones, and thus become transformed at other scales, presents a conceptual framework that may serve the greater goal of transmediating Sesshutei into a spatial sound design. Each of the named Delaunay nodes (dn^1, …, dn^8) thus also represents a loudspeaker position and loudspeaker channel. Various combinations of loudspeakers/loudspeaker channels then define the various zones, for which many of the zones operate across the internal partitioning of the garden as developed in the Voronoi tessellation. Trajectories can be assigned for sounds to operate within a zone (through a simple 1:1 channel mapping) such that the physical movement of the sounds between loudspeaker channels may also approach the concept of a scaling of the sounds through a defining and re-defining of the acoustic arena of the installation.

But the idea of these auditory zones as derivations from the superimposed Drichlet domains of the garden can also be read as an analogous approach to the juxtaposition of varying scales in Sesshū's *Winter landscape* such that the resultant sound design schema equally approaches auditory scale as an opportunity to manipulate, mix and juxtapose acoustic space in a non-linear manner rather than fix it within the confines of a purely linear strategy.

Figure 37: Pd data-structures from audio recordings of Sesshutei

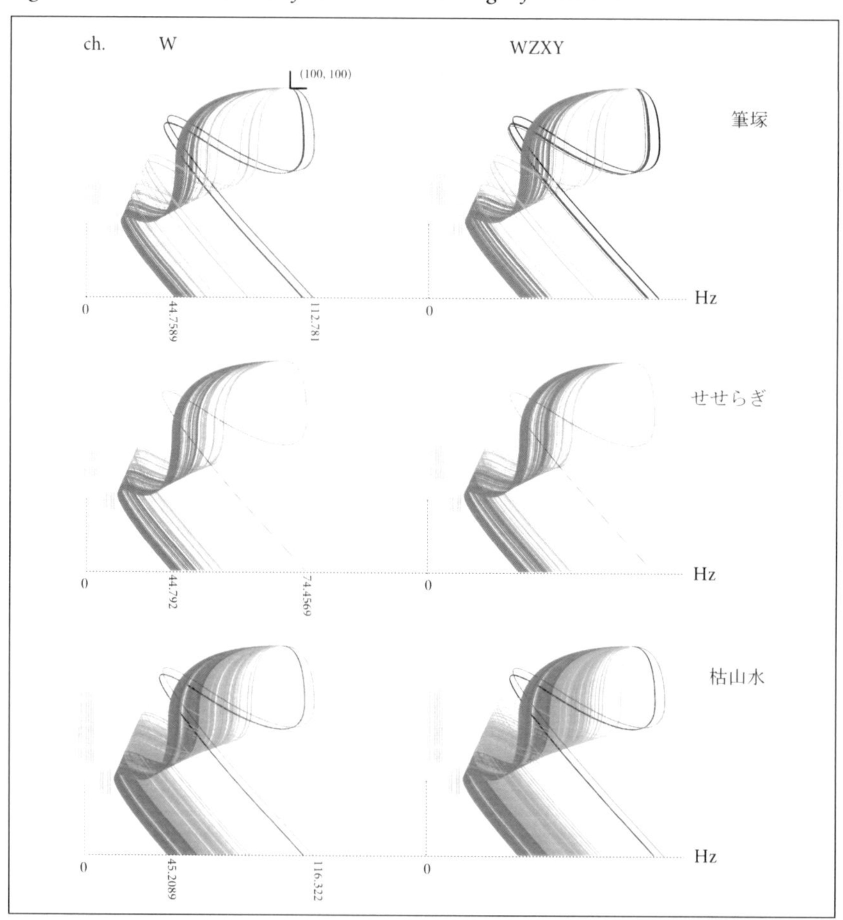

In addition though to utilizing a spatial transmediation of the compositional conditions of the visual landscape within Sesshutei, auditory elements within the soundscape were also mapped and usurped. Unlike its Muromachi contemporary at Ryōan-ji, Sesshutei contains two distinct water features that subtly produce sound sources that activate parts of the garden as embedded acoustic arenas. A small watercourse feeds *shinjiike*, and at the point at which it connects to the pond's northern edge, a soft trickling sound permeates the soundscape. This is repeated further beyond the trees that border the back wall of the viewing scene in which a secondary hidden pond provides a constant flow to this small watercourse. These two subtle soundscape features augment the other sounds of the garden coming from bird life, the sounds of human visitors and the occasional pitched and non-pitched interjections of the temple's Buddhist devotional instruments–the *rin*, *mokyugo* and *daiko* as well as sutra chanting by resident monks.

On an initial field trip in early 2008 these auditory conditions of Sesshutei were captured through spatial sound recordings using an ambisonic B-format microphone. The aim was to both sample the soundscape of the garden as well as extract pertinent spatial acoustic data that could be re-mapped to drive processing engines that transformed those captured soundfiles. The acoustic analysis of the captured soundfiles was again developed in the object-oriented language Pure Data (Pd). Building on the approach used in *Acoustic Intersections*, the `bonk~` and `fiddle~` objects were again utilized to obtain information such as discrete pitch (Hz), amplitude (dB), raw attack spectrum (Hz through 11 frequency bands) and envelope attack from soundfile onset (measured in seconds). In addition to this, numerous visualization strategies that mapped the auditory data of the recordings via Pd's native data-structure capabilities were also explored.

One example of a Pd data-structure is shown in Figure 37. Here, three sites at Sesshutei were recorded: the *fudezuka* (筆塚: a 'burial' site of Sesshū's paint brushes), *seseragi* (せせらぎ: a small babbling stream) and the rock garden (枯山水). The data-structures map the midi note number (0-127) and cooked pitch (Hz) content of the soundfiles through a fixed point (100, 100) via the complete 4-channel ambisonic B-format (WZXY) as well as only through the omni channel (W). As is visually apparent within the data-structure, the variety between the sites is revealed in the consistency and density of sound events. The color of the data-structures maps the number and variety of cooked pitches contained within the soundfiles. The location of the *fudezuka* at the extreme rear of the garden contrasts greatly in pitch material to the content of the auditory experience at the veranda overlooking the rock garden (枯山水). Because of the proximity to the ritualistic chanting and Buddhist instruments of the temple as well as visitor traffic and bird life, the content of the data-structures here in terms of extractable midi notes is far higher and diverse. The *seseragi* presents an intermediatory case. Located half way between the rock garden and the *fudezuka* and directly to the north of *shinjiike*, the *seserag* is a feature that complements both the visual appeal of the rock garden and the temple and the rear heavy tree cover at the *fudezuka* and nearby *karetaki*. The data-structure of the *seseragi* reveals the predictably narrow bandwidth of the sound of slow moving water, and thus a strong concentration around a single pitch that is visible in the data-structure as blue bands. This is in contrast to the rock garden mapping from the veranda in which three distinct midi notes are detectable in the data-structure as dark blue, light green, and light blue bands. The data-structures thus not only provide qualitative visual representation between captured soundfiles of Sesshutei, but also as a means for quantitative interrogation, or for piping real-time information as an output. Using the Voronoi nodes as the capture points for the site recordings, an extensive library of soundfiles was generated. The soundfiles were then variously transformed in their spectral content using numerous objects of the FFTease library of Eric Lyon and Christopher Penrose.[27] The live-processed soundfiles were recorded in real-time so that a timbre library would compliment the sound spatialization schema already developed.

Implementation

What has been apparent in this study of Sesshutei is that the role of the Japanese garden designer or *niwashi* is traditionally centered on considering how to covalence all the available physical elements of the landscape to produce a particular rarefied landscape experience. The conceptual framework for the sound installation *Sesshutei as a spatial model* then was to focus discretely on what the role of an *oto-niwashi* (that is, a sound-gardener) might be. How might an *oto-niwashi* construct particular soundscape experiences that are a function of same types of manipulation and subtle transformation of scale that occur at Sesshutei?

Though I have eschewed the development of the practice of an *oto-niwashi* through the codified channels of traditional music notation, for example, as in a similar manner to the approach Cage used to produce the score of *Ryoanji*, I considered that a *form schema* for structural articulation of the sound installation as an important starting element. The constructed form schema shown in Figure 38 describes a number of the intended parameters of *Sesshutei as a spatial model*, many of which became qualified in real-time during the performance rather than fixed within the prescriptions of the schema as a playback directive. The form schema describes the unfolding of the work in 5 sections (*A*, *B*, *C*, *D*, *E*) whose smallest time unit, *a*, is the basis for the proportional relations (in augmentation or diminution) to the other sections, and is derived from a circuit study of proportional distances between the fixed sound sources within Sesshutei.

The schema also nominates trajectory envelopes for sections (that is, panning information between channels of a zone), a soundfile naming convention, and start/end times for soundfiles (automated mode). The final manifestation of the schema into the performance of Sesshutei as a spatial model is achieved through the concept of "sound planting" or *oto wo tsueru*. There are a certain number of soundfiles that are associated with each of the five sections of the work. These groupings were constructed through linking characterizations of timbre, rhythm and pitch within the files in an effort to construct diverse sound ecologies for each section.

27 Eric Lyon, "Spectral Tuning," in *International Computer Music Conference Proceedings* (Miami: ICMC, 2004), 375-377.

Figure 38: Form Schema for Sesshutei as a spatial model

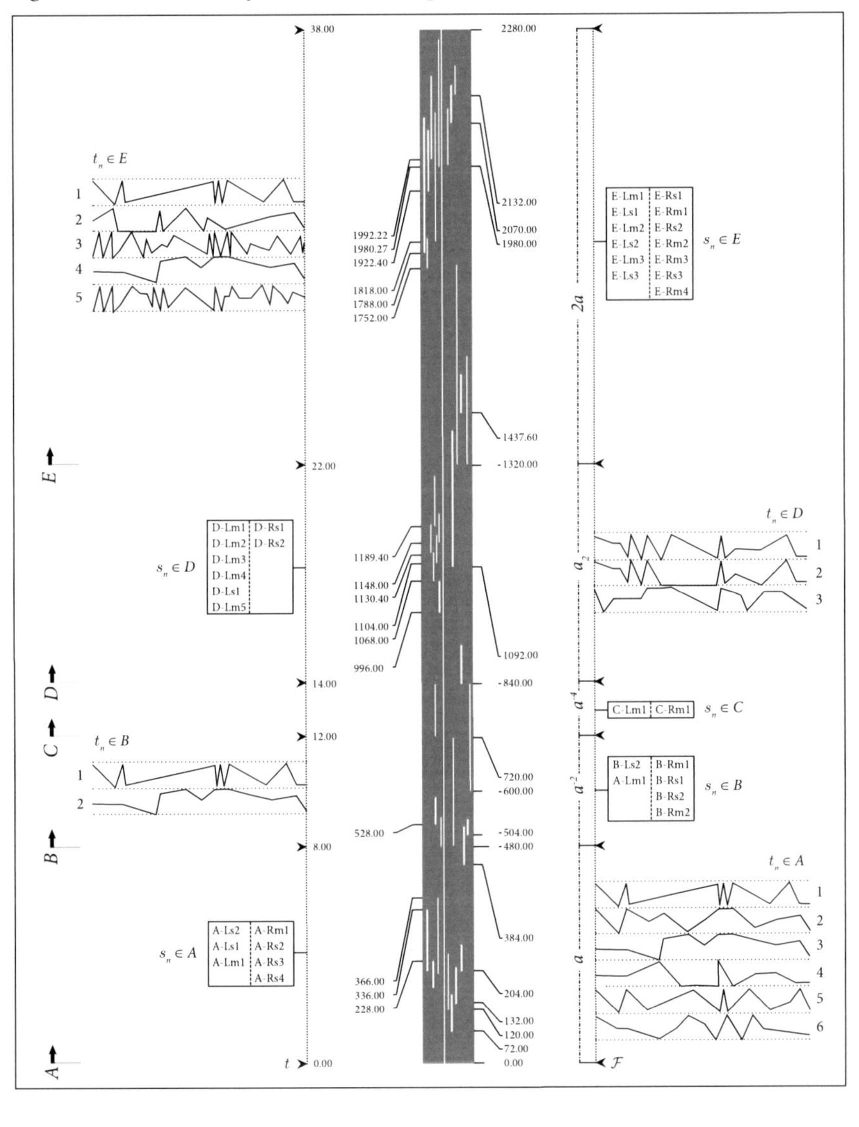

Using Pd again as the playback and spatialization program, messages are send via a TCP socket from a bash shell in the following form:

index-#, soundfile_name, zone-#, env-#;

Soundfiles from the timbre library are called up in real-time and simultaneously assigned a zone number between one and eight that corresponds to one of the eight named auditory zones of the derived Voronoi tessellation.

The envelope number corresponds to one of the available trajectories designated in the form schema (numbered sequentially from 1) according to sections, as well as an additional env-0 command that generates an Ikeda attractor (chaotic process) to govern the trajectory. The speed of the trajectory is automatically scaled according to the length of the soundfile in question, or invoked according to the optional index number. Real-time feedback on the number of soundfiles playing, active Voronoi zones and used channels, dac output faders and event time are displayed using Pd's standard GUI objects, though the inclusion of a TCP network socket and the use of the command line is envisaged as a means to operate the environment remotely, or in Pd's no-GUI mode.

The first performance of *Sesshutei as a spatial model* was presented as a collaborative installation with Yokohama based loudspeaker manufacturer, Taguchi Co. The company creates numerous innovative boutique loudspeakers whose cabinet designs often feature sculptural references to traditional Japanese garden elements such as lanterns and pagodas (Figure 39).

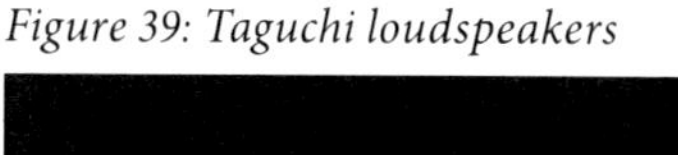

Figure 39: Taguchi loudspeakers

Sesshutei as a spatial model became a collaboration in which a range of their loudspeakers was used in both an indoor and outdoor setting at the Akiyoshidai International Art Village (AIAV). The loudspeakers were used for both an indoor concert in the AIAV Hall as well as one that occurred on the front lawn—both of which utilized eight channels paired to eight subwoofers. In the concert that featured on the front lawn (Figure 40), the loudspeakers were spatially organized in a manner congruent with the dimensions of the original Sesshutei garden. As a response to the modernist Arata Isozaki designed AIAV complex, Taguchi's minimalist CMX-1312/3801 tweeter/subwoofer were used in an effort to homogenize both the visual and acoustic potential of the listening area. The installation sought to synthesize then the elements of the extant landscape in which Isozaki's architectonic gesture and formal materiality provide a controlled geometric balance to the natural sloping topography of the lawn.

A performance area located off the lawn was used for the presentation of a 60 min twilight concert using the developed Pd engine and spatialization patch. The following night, the work was presented within the AIAV Hall using a different array of loudspeakers, many of which operate through a 360° projection of sound. To take advantage of the Hall's irregular floor plan and two levels of balconies meant spreading the 8 channels throughout the space so that sound sources and the panning within zones and between channels could operate vertically as well as horizontally and become a prominent feature of the work. For the performance, Taguchi's more sculptural loudspeakers were employed in an effort to transport and ultimately auralize those recognizable elements of the Japanese garden within the context of the concert hall.

Figure 40: Installation on AIAV lawn

The underlying structural goal of *Sesshutei as a spatial model* then has been to explore the notion of transmediation through a similar appropriation of scale manipulation that is used at the original Sesshutei garden. This approach has produced a spatial sound installation that treats sound design, and Schafer's paradigm of soundscape composition, as a means for developing an insight into what might analogous auditory conditions rely on and how such conditions might be designed or constructed in real-time.

But *Sesshutei as a spatial model* has also attempted to explore the role of artistic works in providing insights into the Japanese garden. In this case, the process of creating *Sesshutei as a spatial model* has revealed how Sesshū's notions about scale and the manipulation of scale he masterfully exploited were not limited to the rice-paper canvas. The correlations between viewer expectations and the sense of scalar juxtapositions and exploitations of the picture space are evident not only in works such as *Winter landscape* but equally within the garden at Sesshutei. Sesshū's work then points towards the notion that space was something that can be constructed *through* the relationship of parts to the whole, of elements and materialities in relation to the void rather than fixed within a container of unchanging metrics. The garden at Sesshutei follows those same 'rules' of his landscape paintings in which the averting and concealing of a common metric to bind the elements under a single domain enables a dynamic and unpredictable rhythm and structure to emerge in the work. This spatial predilection thus became too the basis for *Sesshutei as a spatial model*, where auditory scale and the manipulation, juxtaposition and varied trajectories of sounds are used as a means to produce an equally dynamic acoustic space that seeks to mirror the dimensionless qualities of the landscapes envisaged by Sesshū. Thus just as Sesshū was involved in both the painting of mythical landscapes and then the physical forming of such landscapes through gardening (a type of transmediation process in and of itself), my own artistic approach to the design and implementation of a sound installation has sought to build on this paradigm. In this sense, the process of transmediation provides itself as a valuable tool for which the Japanese garden, as a design space, can be thought of as a set of spatial predilections. Uncovering these relationships and the guiding principles that assist in bringing the separate parts of the garden into a whole that exceeds what can be readily seen, is often achieved through the most basic definition of transmediation—usurping one sign system for use in another sign system. As a secondary, yet closely linked and connected sign system to the picture space of a Japanese garden, auditory design has enabled here the garden to operate within an enclosed and focused medium. That the Japanese garden might be conceived within a purely auditory sphere or context, and thus the role of the gardener as one whom *plants sounds* to build autonomous gardens became the impetus for the final sound installation I will discuss in this chapter, *Shin-Ryōan-ji*.

新龍安時 (Shin-Ryōan-ji)

Cage's precedent for translating the *kare-sansui* at Ryōan-ji into a musical work was driven through both a desire for a musical representation, or analogue to the garden, as well as means for sustaining his exploration of composition as an act that represented the myriad processes within nature. The purpose of music, as Cage famously argued it, was in "imitating nature in its manner of operation."[28] In terms of the scope of an *oto-niwashi*, Cage's ideas about the nature of musical systems also seem an important catalyst to the concerns of sound gardening given the composer's assertion that music *is* ecological. Both this contention, and the nature of his approach to transmediating Ryōan-ji have provided a stepping off point for my own implementation of sound design as a gardening discipline. In *Sesshutei as a spatial model* the sound gardener was positioned as a real-time composer of sound-space, and one who in a very specific and goal-oriented fashion *crafted* the dimensions and illusions to auditory scale during the unfolding of the work. But what might the implications be for simply approaching sound gardening as a design of a contained range for which sounds may or may not unfold? What if the sounds and their characteristics were simply a function of a *potential*, that is, their dynamics, rhythmic fluctuation, texture and timbre were controlled not in a linear or direct fashion by the sound gardener, but statistically or randomly by a set of rules? This model of approach perhaps still retains the central idea of the *oto-niwashi*, though now the approach to sounds is one that is one step removed in the sense that the sounds are designed to operate within a *continuum* of potential states rather than exist as fixed objects as in *Sesshutei as a spatial model*. In this paradigm, sounds may be initiated in one state, then transform or evolve over the course of a period of time for which the sound garden grows. The sound gardener then is one who decides on the *variability* of the sounds and the dimensions of the constraints to the change, but is one who is only statistically aware of the course and direction that the sound design may take. Perhaps this similarly approaches Linton's notion of how a landscape's *vividness* and *variety* seem at particular times to reach a zenith that may not always of been readily predictable or foreseeable.

As a secondary response (after *Re-Ryoanji*) to the *kare-sansui* in Kyoto, *Shin-Ryōan-ji* (trans. new Ryōan-ji) seeks to somewhat again divert the path of Cage's precedent by usurping a move from a model of the composer or *oto-niwashi* as agent to the notion of the palette or collection of designed sounds themselves as autonomous agents. As such, *Shin-Ryōan-ji* developed as a multi-channel sound installation that both maps the spatial proclivities of the original Kyoto site while attempting to construct a deeper conceptual framework by further exploring and interrogating the constituent principles of what *oto-niwashi* sound design is consistent with. As such, the premise of the installation relies on a digitally curated ecology of sounds that becomes distributed and periodically 'fed' to produce a *digital garden*. In this sense the installation takes cues not only from Cage and his argument for a synthesis between the operations in nature and musical composition, but also from the concept of the feeding of environmental invariants to produce a sound design that was explored in *Acoustic Intersections*. The work then is both homage to Cage as well as yet another departure point for exploring how the notion of transmediation of auditory spatiality in Japanese garden design may enable a continually evolving path for exploration.

28 James Pritchett, *The Music of John Cage* (Cambridge: Cambridge University Press, 1993), 97.

Figure 41: Shin-Ryōan-ji, Japan Foundation, Sydney

A Digital Sound Ecology

Given that the most striking element of the original *kare-sansui* at the Ryōan-ji temple lies in the geometry of the site, and the various perspectives obtained through viewing from the temple veranda, the representation of rock groupings in *Shin-Ryōan-ji* retains the geometry of the original site, though is scaled accordingly to the dimensions of the original presentation space (Japan Foundation Sydney, see Figure 41 and 42). In both an attempt to abstract the perceived rhythmic structure of the grouping, and provide a means to deliver an acoustic ecology into the installation site, open, black cylindrical representations of the rocks were designed to allow for various sized loudspeakers to be placed within the cylinders, as well as to facilitate a method by which sounds might be *tuned* to the site through cylinder-sound-source mappings. The sense of the installation as a *digital sound ecology* was produced through a use of open-source sound synthesis methods in Pd that were driven by the program's implementation of cellular automata modeling (that is, via the `mlife` object).

Cellular Automata (CA) systems are dynamic in the sense that they are discrete in terms of space and time. They function through a collection of cells, each of which obeys a common rule among all cells and each of which has a small number of states (usually on = 1 or off = 0). The rules may vary greatly between CA models with John Conway's Game of Life perhaps the most famous,[29] but as Andrew Iiachinski notes, all CA models tend towards containing 5 generic characteristics that can be described as:

- **Discrete lattice of cells**: the system substrate consists of one-, two- or three-dimensional lattice of cells.
- **Homogeneity**: all cells are equivalent.
- **Discrete states**: each cell takes on one of a finite number of possible discrete states.
- **Local interactions**: each cell interacts only with cells in its local neighborhood.
- **Discrete dynamics**: at each discrete unit time, each cell updates its current state according to a transition rule taking into account the states of cells in the neighborhood.[30]

Figure 42: Shin-Ryōan-ji

29 Richard K. Guy, "Conway's Prime Producing Machine," *Mathematics Magazine* 56/1 (January 1983): 26–33.
30 Andrew Iiachinski, *Cellular Automata, a discrete universe* (London: World Scientific, 2001), 5.

Mathematically, a CA can be described as a finite-dimensional lattice in which each site value is restricted to a (usually) small set of integers such that $Z_k = \{0, 1, \ldots, k-1\}$. As Erica Jen outlines,[31] a general form of a one-dimensional CA is given by:

$$x_i^{t+1} = f(x_{i-r}^t, \ldots, x_i^t, \ldots x_{i+r}^t),$$
$$f: Z_k^{2r+1} \longrightarrow Z_k$$

Here x is a cell and is indexed via its site location, i at time t, while f is the "rule" that governs the fitness of each cell in successive generations. A non-negative integer is assigned to r that specifies the radius of the rule f. The installation design of *Shin-Ryōan-ji* used a simple CA model implemented in Pd such that fifteen fundamental sound engines, (S_n), were driven by a lattice, Z_{15}, via a Pd `mlife` object. Over the course of a vast number of evolutionary iterations, with each evolutionary period (t) lasting 390 seconds, the current cell state messages of $x_i \in Z_{15}$ of 1 (alive) or 0 (dead) were send to S_n as on/off messages. This approach thus enabled the state of the CA to govern the fluctuation of the fifteen sounds in terms of texture, polyphony and rhythm—hence the allusion to a *digital sound garden*. The design of the sounds of the garden was kept rather simplistic. Simple oscillators, phasors, AM synthesis, signal vector swapping and sample playback combined with filtering was used to construct a seemingly complex and dynamic soundscape that seeks to approach a transmediation of the original *kare-sansui* as an *organic* entity (see Table 7). In effect, fluctuations between the fifteen sound classes were articulated through variations in pitch content and timbre spectrum so as to appropriate an auditory analogue to the fifteen rocks of Ryōan-ji (see Figure 43).

Table 7: Descriptions of Pd Sound engines in Shin-Ryōan-ji

Sound Engine	Description
S_1 = blockswap~	swap upper and lower half of any given signal-vector: break at 32 samples (swap) and 64 samples (signal-vector)
S_2 = noish~ 25.92	band limited noise in which bandwidth is controlled by a drawing rate n (Hz)
S_3 = beatify~	audio amplitude modulator (ADSR + loop size)
S_4 = quantize~	quantize a given signal with variable step number

31 Erica Jen, "Aperiodicity in One-dimensional Cellular Automata," in *Cellular Automata*, ed. Howard Gutowiz (Amsterdam: Elsevier Science Publishers, 1990), 3.

S_5 = swap~	byte swap a 16 bit signal; first conversion swaps the upper-lower byte
S_6 = blockmirror~	playback signal vector reversed in time (break at 64 samples—signal vector)
S_7 = disto~	distort signal with high and low pass filters
S_8 = phasor~	sawtooth generator
S_9 = beatify~_2	audio amplitude modulator (ADSR + loop size)
S_{10} = beatify~_3	audio amplitude modulator (ADSR + loop size)
S_{11} = readsf~	read a soundfile
S_{12} = readsf~_2	read a soundfile
S_{13} = sampler	store a soundfile in a table and playback utilizing variations in pitch, sample rate, length
S_{14} = sampler_2	store a soundfile in a table and playback utilizing variations in pitch, sample rate, length
S_{15} = ring modulator + qmult~	multiplying a complex tone by a sinusoid; multiplication of 2 quaternion signals

Each of the fifteen sounds, perhaps like the original rocks of the garden, were designed to be highly contrastive, seemingly angled through varying rhythmic articulations so that above all, they remained distinct from each other. This sonic transmediation of the original garden's rocks was also curated in such a fashion as to allow for a maximum amount of sonic indeterminacy throughout the life and evolutionary output of the fifteen sound engines. By this I mean that each of the fifteen sounds, when initialized, had the potential to radically change within its various evolutionary steps. This was managed through periodically controlled parameter messages sent to the `mlife` object in which random seeding and neighborhood values (that is, assigning 0/1 values for each cell) were regularly updated in the CA. This means that the general condition of each sound engine is a temporal function of on/off (0/1) messages from the `mlife` CA, and is refreshed through updates send out every 390s. But in addition to his approach, a enlarging of the concept of the transmediation of Ryōan-ji was put into effect through the design of the potential for fluctuation *within* each of the sound engines. This is accomplished through randomly controlled internal parameter manipulations within each sound engine. Such a model then allows for a large variation in the timbre, pitch, and general behavior of each sound engine as well as dynamism in the articulation of the sound-space within the site. When combined with the other random messages sent to the `mlife` CA to control the state of cells, individual cell values, or the neighborhood values, an auditory environment arises in which sounds grow, evolve and atrophy across the fifteen channels in a range of temporal states that manifest both the macro and micro unfolding of time, and are governed by the population flux densities of the digital cell culture. Additionally, there is also a range of rhythmic variation among the sound engines in an attempt to balance both long flowing sounds, to sounds of sharp attacks or those that are repetitious in nature. The larger rhythmic continuity is duly controlled by the on/off messages of the `mlife` CA, which means that long periods of stasis of only a few sounds active can be abruptly interrupted by an appearance of all fifteen engines.

Figure 43: CA schema and Shin-Ryōan-ji plan

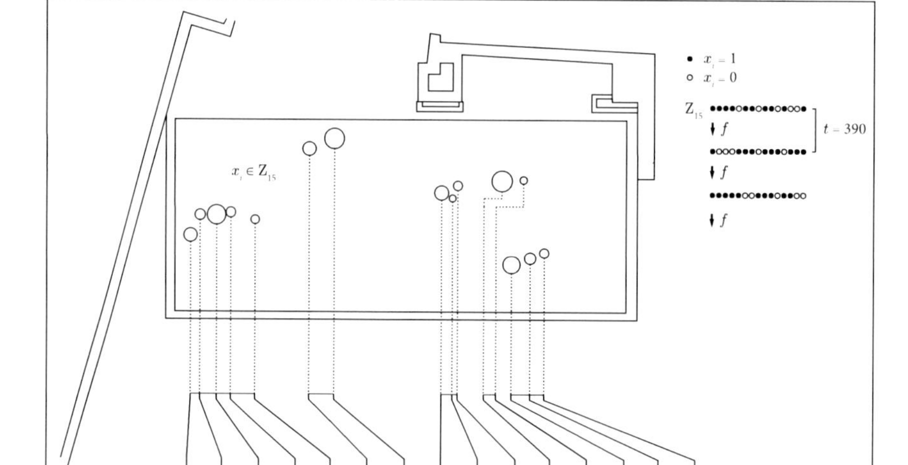

As a design space, *Shin-Ryōan-ji* explores the notion of a group planting of sounds to which the ebb and flow of auditory information is indeterminate, or chaotic, yet the spatial locations fixed. As an analogue to the nature of gardening, the installation represents an exploration of the process of cultivation. The work is a function of both a mapping of spatiality within Japanese garden design, and an analogous, auditory dimension in which sounds mimic the life cycle of organic entities. The potential for a varied, yet meditative soundscape to evolve within the installation site approaches the seasonal nature of Japanese garden design, in which the cyclic effects of the natural world have a profound effect on both the soundscape and visual elements within a garden. More specifically, the influence of Cage's work on the design of *Shin-Ryōan-ji* comes from the desire to explore indeterminacy as a real-time process, and provide a point of continuation of Cage's musical aesthetics.

But in regards to the concept of the *oto-niwashi* or sound gardener, *Shin-Ryōan-ji* departs from the model explored in *Sesshutei as a spatial model* in that rather than the gardener as agent and composer of the exact and specific dimensions of sound-space, now sounds themselves play a more pivotal role. In *Shin-Ryōan-ji* the sounds themselves become agents and their characteristics and temporality is not a readily known fact of the gardener. Rather, the gardener has a fundamental statistical knowledge of the range to which they may grow, and the role is simply to put the sounds in play and to experience the design as it unfolds. Sound design then in the work has been reconsidered: instead of having an intimate knowledge of the desired qualities and structure of the soundscape to be generated, the *oto-niwashi* has a more generalized role that simply sets the outer *limits* for what may transpire, in effect eliminating a particular desirability for one determined outcome to transpire over another. In this model of sound design, the range of the possibilities of the eventual soundscape is the underlying aesthetic motivation.

Such a framework perhaps approaches Cage's musical aesthetics in a rather close manner. Cage explained that his relationship to the thematic materials of composition were one in which the composer's directive is to discover "means to let sounds be themselves rather than vehicles for man-made theories or expressions of human sentiments."[32] The approach towards the sound design within *Shin-Ryōan-ji* then is perhaps not only a model for the transmediation of the original *kare-sansui*, but similarly another stepping-off point in a continued exploration of Cage's relevance towards the usefulness of the Japanese garden as a spatial paradigm. But the integration of the `mlife` CA in *Shin-Ryōan-ji* has also pushed the exploration of the approach of an *oto-niwashi* more firmly into Cage's aesthetic realm of the embrace of indeterminacy. By utilizing a CA driven through random number generators to evoke fluctuations to the unfolding and sound-space mimics somewhat Cage's notion that

"a sound does not view itself as thought, as ought, as needing another sound for its elucidation, as etc.; it has no time for any consideration—it is occupied with the performance of its characteristics: before it has died away it must have made perfectly exact its frequency, its loudness, its length, its overtone structure, the precise morphology of these and of itself."[33]

Shin-Ryōan-ji, as a digital garden, can perhaps be described as such, in which the collective of individual sounds function independently of one another, yet are controlled from a centralized cell culture that itself is highly contingent on cell neighborhood states and their influence on the unfolding evolution of the culture. But perhaps this also serves somewhat as an analogy to the four sound design projects presented thus far, *Re-Ryoanji*, *Acoustic Intersections*, *Sesshutei as a spatial model*, and *Shin-Ryōan-ji*. For each, an experiment into the spatial attributes of the Japanese garden and how

32 Cage, *Silence*, 10.
33 Cage, *Silence*, 14.

these attributes may enable approaches to soundscape design to emerge while at the same time enabling an interrogation and exploration of the role of an *oto-niwashi*. Each project is a self-contained one, though they are inevitably linked through the influence and questioning of Cage's aesthetics and Cage's *Ryoanji* precedent.

On Gardening with Sounds

The four sound design works explored in this chapter have not only presented themselves as experiments in an ongoing homage to Cage and the perceived role of an *oto-niwashi*, but have equally used the broader platform of soundscape composition to position creative work as a method for ecological and acoustic analysis. Indeed, from the beginnings of the acoustic ecology movement, Schafer, Truax and Westerkamp have argued on the importance of creative practice and the role of soundscape composition in producing new knowledge about place and the relationship between inhabitants and their sounding environment. This position is evident in the broad nature of Schafer's original definition of *soundscape*. In the early 1970s he defined it as:

"Any acoustic field of study. We may speak of a musical composition as a soundscape, or a radio program as a soundscape or an acoustic environment as a soundscape".[34]

From the beginnings of *The World Soundscape Project* in the early 1970s as a vehicle intended to foster a greater public awareness of noise pollution in the built environment, the role of soundscape composition is one in which auditory awareness for the listener becomes a driving factor. In fact, the inheritance of the advances in digital studio techniques from technological developments in the 70s and 80s have only strengthened the importance today within the field of acoustic ecology of electroacoustic compositional methods for generating soundscape compositions and soundscape simulations. Truax has noted of soundscape compositions that: "the intent is always to reveal a deeper level of signification inherent within the sound and to invoke the listener's semantic associations without obliterating the sound's recognizability."[35] Though Truax, Schafer and Westerkamp come to the discipline of acoustic ecology as composers, the aesthetics of soundscape compositions need not adhere to the traditional aesthetics of musical composition. Given that abstract instrumental (that is, non-text based) music is always self-referential, a usurpation of sounds from a non-musical acoustic environment for use in a musical one will by traditional musical aesthetics, necessitate an examination on the relation of sound objects to each other, rather than to their social or environmental function (that is, the extra-musical or the ecological).

34 Schafer, *The Soundscape*, 7.

35 Barry Truax, "Soundscape Composition," accessed January 7, 2014, http://www.sfu.ca/~truax/scomp.html

Here, the aesthetics of soundscape composition naturally diverges from the traditional. Rather than the sole focus on thematic materials and the inter-relationship between parts, the aesthetics of soundscape composition presents to us what Angus Carlyle notes as an argument for understanding of our place within the environment:

"The incorporation of the human into the definition of the landscape—and, by extension, into that of the soundscape—focuses attention directly into questions of perspective".[36]

The original emphasis of Truax and Schafer that *any* sonic environment, be it a man-modified, designed, simulated or natural environment, is a communicational system implies that the role of the listener must become activated. In taking such an approach, Truax positions acoustic ecology and soundscape composition as an understanding of how sound, when delivered to our auditory facilities, travels in a system of information exchange (through the delivery of environmental invariants), rather than simply a transfer of energy. In considering the information processing approach Truax argues that the value of context is extremely important:

"The exchange of information is highly dependent on context, whereas the transfer of energy is not [. . .] In a communicational approach, context is essential for understanding the meaning of any message, including sound".[37]

Ultimately, for eco-structuralist sound design to be a successful tool in understanding the acoustic environment, there must be a method to which acoustic design is directly informed from Truax's communication model of a soundscape.

Indeed, throughout my explorations of the Japanese garden, the concept of the *oto-niwashi* has arisen as a means to better understand, from the stance of the gardener, what is actually to be, or being communicated through the sound design. Though each of the four sound design projects, *Re-Ryoanji, Acoustic Intersections, Sesshutei as a spatial model,* and *Shin-Ryōan-ji* have developed in somewhat of a direct response to Cage's musical precedent *Ryoanji*, the goal of each as a transmediation of the Japanese garden has been to interrogate the structural principles of the sampled gardens. How these underlying structures might be re-combined, re-sampled, or re-configured so as to produce particularly meaningful auditory experiences is perhaps the ultimate goal of the *oto-niwashi*. In this sense, the role of the *oto-niwashi* equally approaches the notion that soundscape composition is a means for ecological awareness through the lens of acoustic knowledge.

36 Angus Carlyle, "Like Quails Clucking: Landscape, Soundscape and a Noisy Future?" in *The East Meets the West in Acoustic Ecology*, eds. Tadahiko Imada, Kozo Hiramatsu and Keiko Torigoe (Hirosaki: JASE & HIMC, 2006), 113.
37 Truax, *Acoustic Communication*, 11.

But rather than seeking an obvious or low-level adaptation of Cage's compositional methodology, though at the same time intrigued by the concept of how a sound garden might be otherwise composed, each of the four sound design projects explores the notion of *gardening with sounds*. By being privy to the role of the gardener and thus stepping into the role of designer, allows for a usurpation of those spatial proclivities of the Japanese garden in a meaningful way. What is being communicated then within all the projects is that which also drives the original structures of the sample gardens: a balance of forces of flux and stasis, a mix and play of scales, and a realization of the organic nature of a garden is a function of natural form *and* the hand of the gardener.

The adaptation of analytical methods such as Voronoi tessellation, multichannel audio recordings and data visualization have been combined with other generative approaches such as CA, sound synthesis, acoustic convolution, and sound spatialization in an effort to produce a creative *toolbox*. If the traditional Japanese gardener is guided by Master-apprentice training and the breadth of knowledge contained within the extant gardening treatises, then the explorations in creative practice and soundscape composition outlined here have duly sought to generate a contemporary toolbox-as-treatise. Rather than copying the traditional Japanese garden and seeking to replicate it purely within the auditory realm, a toolbox-as-treatise is an approach that combines analysis and synthesis as a means to usurp the sign system constructed by the Japanese garden as a viable sign system to inform new sound designs and soundscape compositions. Perhaps this is the strength of the collected techniques of the toolbox: that their demarcation of the proclivities of the Japanese garden as a site for landscape experiences enables numerous types of transmediations of spatial auditory experiences to emerge. In this sense, the methods within the toolbox offer ways of constructing soundscape architectures born through considerations about site and scale, nature and order and the role of auditory content as a connective tissue.

But the transmediation of spatial information from the physical tangible aspects of a garden as a landscape, into a garden as a soundscape, also means that the consideration of the produced visual experience still remains an import channel. This suggests that gardening with sounds, as *oto-niwashi* may not be the only avenue in which to pursue the Japanese garden as a paradigm for transmediation of its auditory qualities. As the Japanese garden is an integrated system, for which the designed multi-sensory experience becomes a mediated synthesis between the tangible conditions of landscape and the intangible conditions of soundscape, an emphasis on the three dimensional visualization of these interrelations may provide yet another creative opening for investigating the garden as a spatial paradigm.

Architectonic Form Finding

The impact of digital modeling and systems for CAD (computer-aided design) within all fields of design practice, but in particular, in architecture, is a result of what Kyle Davy identifies as the "exponential growth in computational performance, digital memory and storage, digital communication, and the internet."[38] The impact within architecture as particularly noticeable with the explosion of 90s *blobitecture* and the movement towards a rejection of ruled and straight surfaces in favor of slinky curvilinear forms. As John Waters notes, "these post-war designers and architects drew their inspiration from nature, and manifested their anti-machine impulses in designs for buildings and objects that were, if not strictly blobs, certainly more biological than the prevailing style."[39] Typical architectural projects from leaders in the field such as Frank Gehry, Zaha Hadid, Mark Gulthorpe and Norman Foster attest to the allure that digital and parametric form modeling, existing as flexible relationships rather than fixed structures, have been significant in pushing architecture beyond its own well-worn traditions and theoretical stances. As Davy explains:

"In the past, architects relied on sketches and the occasional handcrafted model as the primary media for expressing and communicating design ideas. Now equipped with a wide array of digital design and modeling technologies, including 3-D design, BIM [building information modeling], energy-use simulators, and quantity take-off and cost databases, architects are transitioning to new mixed-media design environments."[40]

But this new found digital competence has also enabled the discipline of architecture to become more porous and susceptible to outside influences. For architect Marcus Novak the potential of contemporary digital tools to generate virtual environments gives rise to a new type of space in which architecture and music can coexists, the resultant environment termed "archimusic."[41] While Novak's conception may seem novel in its intentions to redefine the very notions of what architecture is, Michael Ostwald has already highlighted the fact that architecture today has been highly influenced by numerous external disciplines.[42] This has arisen through its historical eagerness to seek out myriad theories and extra-architectural knowledge in an effort to expand and push the discipline into new territories. Indeed, this paradigm of knowledge acquisition has also led both Jean-Claude Guédon[43] and Botand Bo-

38 Kyle V. Davy, "The Design Process, catching the third wave," in *Architecture, celebrating the past, designing the future*, ed. Nancy B. Solomon (New York: Visual Reference Publications, 2008,) 161.
39 John Kevin Walters, *Blobitecture: Waveform Architecture and Digital Design* (Glouster, MA: Rockport Publishers Inc., 2003), 12.
40 Walters, *Blobitecture*, 161.
41 Marcus Novak, "Computation and composition" in *Architectural as a translation of music*, ed. Elizabeth Martin (New York: Princeton Architectural Press, 1994), 66.
42 Michael Ostwald, "Architectural Theory Formation through Appropriation," *Architectural Theory Review* 4/2 (1999): 52-70.
43 Jean-Claude Guédon, "Architecture as Transdisciplinary Knowledge," in *Anyplace*, ed. Cynthia C. Davidson (Cambridge: MIT Press, 1995), 88-99.

gnar[44] to argue that architecture has indeed ceased to occupy a finite domain—its very boundaries have dissipated as the definition of what architecture *is* continues to be re-thought. While noted architect Lebbeus Woods had countered this notion by arguing that "over the past few decades, architecture as an idea and practice has increasingly limited its definition of itself," he nonetheless opined that "what is needed desperately today are approaches to architecture that can free its potential to transform our ways of thinking and acting."[45] Novak's concept of "archimusic" then works very much towards this goal, though the concept is one that is intimately connected to the notion of the "dataworld." Here, the act of spatial composition is one that is facilitated through the utilization of algorithms that generate music and architectonic form *simultaneously*. Spatial information then is simply a potential that may become structured and fixed by the designer to resemble sound forms, architectonic forms or a kinetic combination of the two. The conduit of the digital environment thus allows for these structures to be in constant flux, movement or symbiosis:

"Cyberspace is liquid, liquid cyberspace, liquid architecture, liquid cities. Liquid architecture is more than kinetic architecture, robotic architecture, an architecture of fixed parts and variable links. Liquid architecture is an architecture that breathes, pulses and leaps as one form and lands as another."[46]

Novak's argument then is that digital or synthetic environments are qualitatively filled with spatial potential such that navigation within the environment is akin to quantifying its informational content, or forming it into structures. The nature and channels to which this spatial information becomes assigned and *objectified* is not a function of traditional disciplinary boundaries. The inherent fluidity of the synthetic environment allows cross-disciplinary movement, experimentation and an embrace of impermanence and ephemerality—themes not traditionally embraced by the cult of the architectural monument.

As a contemporary to Novak, architect Greg Lynn has also championed 3-D computer modeling and animation as a way to manipulate form outside of the boundaries and perceived restraints of traditional architectural design drafting. Indeed Peter Szalapaj sees Lynn's embracing of digital technologies as a consequence of the rapid progression of technological advancement in the last twenty years such that "the range and extent of CAD-related applications in architectural practice have increased beyond the old-fashioned perception of CAD as merely a production tool."[47] But for Lynn there is also the lure of radicalizing architectural form through real-time

44 Botand Bognar, "Anywhere in Japan," *Architectural Design* 62/11-12 (1992): 70-72.
45 Lebbeus Woods, foreword to *Ambiguous Spaces, NaJa & deOstos*, by Nannette Jackowski and Ricardo de Ostos (New York: Princeton Architectural Press, 2008), 5.
46 Marcus Novak, "Liquid Architectures in Cyberspace," in *Cyberspace: First Steps*, ed. M. Benedikt (Cambridge/London: MIT Press, 1991), 250.
47 Peter Szalapaj, *Contemporary Architecture and the Digital Design Process* (Oxford: Architectural Press, 2005), 4.

software manipulation, which stems from his desire (much like Novak's) to counter architecture's dedication to permanence and the inert:

"By collapsing the plan and elevation into a single three-dimensional form, the computer model encourages architects to break out of orthogonal, right-angled space. Instead they can manipulate all dimensions at once. Space becomes as plastic as a hunk of Silly Putty".[48]

Though the contemporary discourse on the nexus between architecture and music has tended to focus on design methodologies for the translation of one into the other, the impetus for this focus has no doubt arisen from an historical fascination between the two disciplines. What the potential of digital architectural modeling may have for understanding the underlying spatial principles of the Japanese garden and its soundscape qualities may enable a new avenue to open up that neither completely adheres to Novak's compositional environment of "archimusic" nor the contention of Goethe that "architecture is frozen music."

In the following discussion I outline three projects—*NURBS Map I / II*, *Kyu Furukawa Teien Shūsaku* and «*No-stop garden*»—that utilize digital architectural modeling tools as a means to investigate spatial relationships within the two Japanese gardens examined in chapter 3: Kyu Furukawa Teien and Koishikawa Korakuen. NURBS or Non-Uniform Rational B-spline modeling is a technique for generating and manipulating curves, surfaces and architectonic forms within a computer. The three projects discussed were a result of experiments using NURBS techniques for form generation and parametric manipulation in the CAD software Rhinoceros 3-D. Each of the projects here have attempted to use digital architectonic modeling as a means to project spatial information regarding the auditory conditions at the two gardens into a geometric context that represents the information through NURBS surfaces. The deeper premise of the approach being that the spatial information within the Japanese garden produces not only landscape structures but also sound forms. How this inherent relationship might be visualized as an abstracted 3-D object or an abstracted 3-D architectonic map may provide further insights into the fundamental structures of the garden, while at the same time building on the notion of Novak and Lynn of the fluidity of spatial information as a potential compositional resource. There is of course a subjective compositional approach to each of the spatial methodologies I explore, such that the transmediations of the data are relative to each project. For *NURBS Map I / II* and *Kyu Furukawa Teien Shūsaku* the aims of the methodology were to simply represent auditory and acoustic data within an architectural framework for evaluation and comparison. In «*No-stop garden*» the motive for experimentation lies not so much within an analytic framework, but moreover, in an opportunity to further extend the FCA analysis and Hasse diagram of Kyu Furukawa Teien (chapter 3) as a prescriptive device for an exploration into parametric mode-

48 Walters, *Blobitecture*, 67.

ling in Rhinoceros 3-D. But though differences in approach remain slight between projects, the greater commonalities between them rests in the desire to map auditory spatial information from the Japanese garden into visual sound forms that enable a new means for *spatial thinking*.

NURBS MAP I / II

As was discussed in chapter 3, the gardens of Kyu Furukawa Teien and Koishikawa Korakuen in Tokyo were examined using the audio recording techniques and mapping methodologies of acoustic ecology as a way to not only document the sound sources of the gardens, but also to acquire insights into how both gardens construct unique spatial auditory experiences. The recording field trips occurred in the early summer of 2006, with a total of twenty-four multi-channel recordings that captured the spatial auditory invariants within Kyu Furukawa Teien and Koishikawa Korakuen. In addition to these audio recordings, extensive data was gathered that recorded SPL_A (A-weighted sound pressure levels) at various points throughout each garden. A-weighted sound pressure level recording uses a filter curve on lower frequency auditory content in an effort to compensate for the relative loudness perceived by the human ear, which is less sensitive at these frequencies. Given that SPL_A data is only represented as a number measured in decibels (dB) or an average dB of readings within a particular time period, it gives only a more general concept of the conditions within the gardens in terms of the perception of sound source loudness rather than sound source qualities (such as timbre or pitch). But just as there is an element of play within each garden relating to the notion of *soto* and *uchi*, the SPL_A data set obtained did manage to highlight the (predictable) auditory trends between the exterior and interior of the gardens as well as the not so obvious tension and pull of opposites within the variously occurring interior micro-environments.

An initial mapping of the micro environments of each garden into spatial schema highlighted how the ambient SPL_A data provided further insights into the articulation and formation of quiet and transitional zones and spaces of flow within each site. Figure 44 and 45 shows generalized spatial diagrams of Koishikawa Korakuen and Kyu Furukawa Teien. To explore the auditory relationships inherent in the site, each of the designed water features is located within the diagram in relation to their distance from the main entry point of the garden. This linear approach allows for distances between auditory encounters within the garden relative to the obvious first encounter to be assessed. In this first encounter of the movement of the listener from the city and its heterogeneous soundscape into the confines of the garden and its more balanced yet homogenous soundscape presents a paradigm to which the other auditory encounters of the garden can be read as variations or diversions from this structural base.

Figure 44: Spatial schema, Koishikawa Korakuen

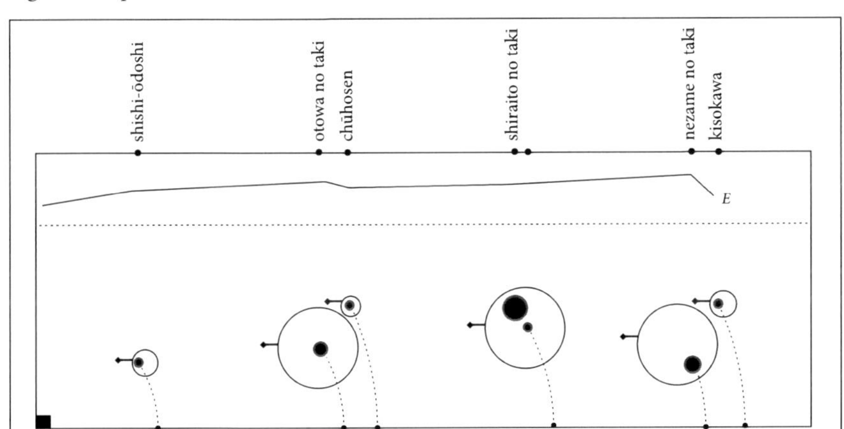

Figure 45: Spatial schema, Kyu Furukawa Teien

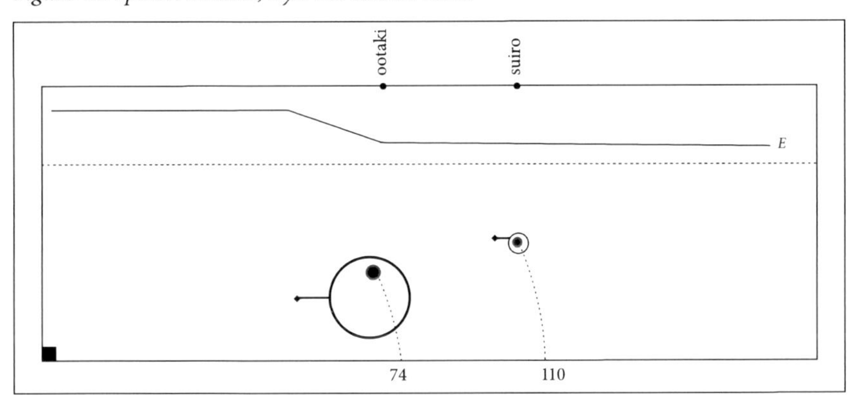

Within the schema, a circle represents each of the encountered micro environments of the gardens and corresponds to their relative size. Within each circle is an indication of the number of primary sound sources (filled circles) of the site, and its relative dB level (size of the circle). Small arms extend off the circumference to indicate relative pitch value (in Hz) for the micro environment. Additionally, each site's relative elevation and generalized topography is also included. These spatial diagrams are an attempt to represent the relationships between each of the designed sound elements that define the micro environments of the gardens. What become obvious from them is the function and importance of designed water features to act as points of spatial articulation within the gardens.

Figure 46: SPL_A recording point at Koishikawa Korakuen and Kyu Furukawa Teien

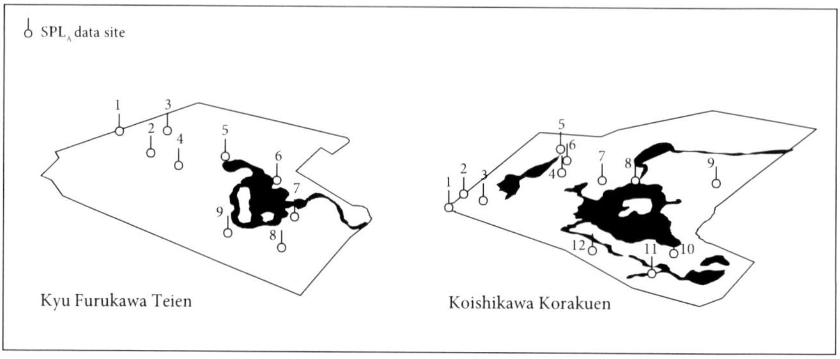

Table 8: SPL_A data from Kyu Furukawa Teien and Koishikawa Korakuen

Kyu Furukawa Teien		Koishikawa Korakuen	
data site #	SPL_A value	data site #	SPL_A value
1	93.0	1	66.5
2	81.1	2	56.1
3	63.0	3	51.7
4	53.8	4	60.1
5	62.9	5	50.6
6	42.4	6	50.0
7	52.6	7	59.3
8	77.2	8	63.0
9	45.9	9	64.0
		10	56.8
		11	59.2
		12	51.7

But further to this two-dimensional representation of pertinent site information, was a desire to project and visualize the complete SPL_A data sets from across each garden as three-dimensional representations. Thus, rather than simply representing this information through data plots or in tabular form, a 3-D visualization technique

Figure 47: NURBS Map I (Koishikawa Korakuen)

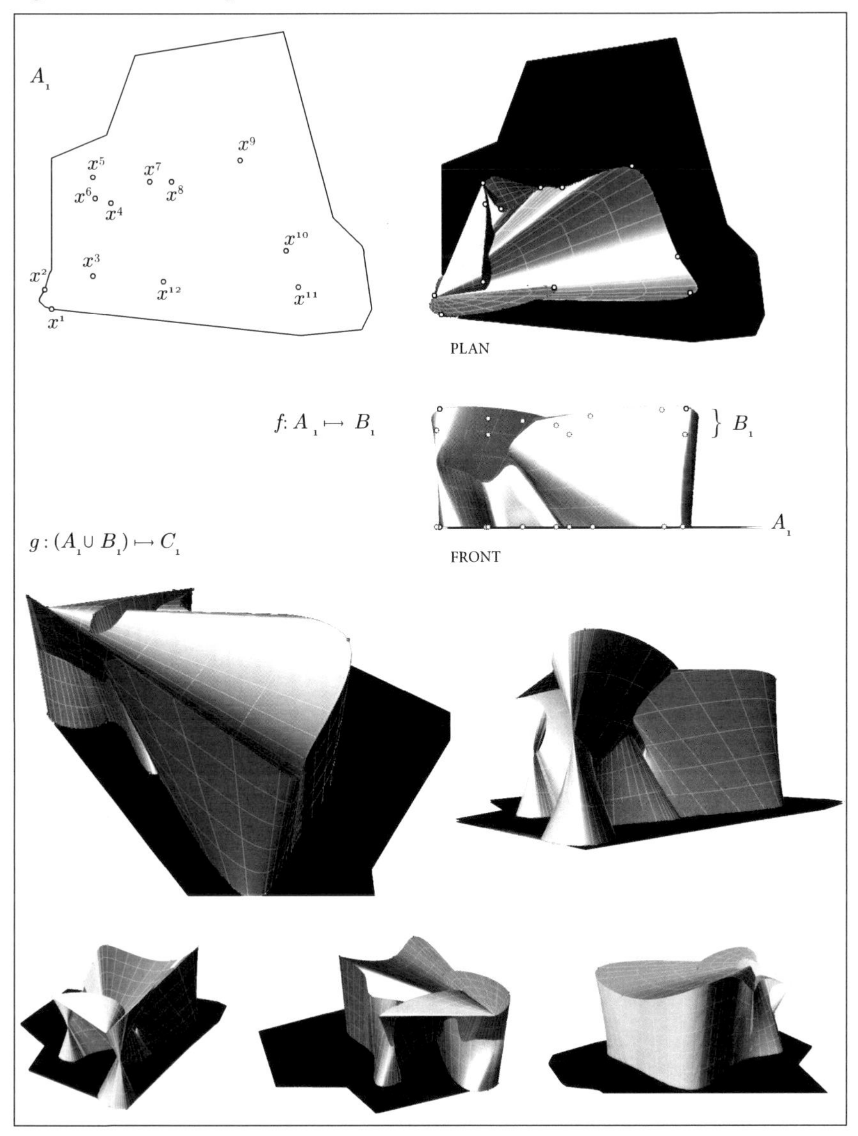

using NURBS architectonic form generation was devised in an effort to more readily and qualitatively highlight these interrelations. Taking the data set of the ambient SPL_A levels recorded from both gardens (see Figure 46 and Table 8) was the starting point for a process for the creation of three-dimensional architectonic forms. This process visualizes the relationship between a theoretical SPL_A 0dB level called set *A*, and a corresponding ambient SPL_A value named *B* (from values in Table 8), and in doing so, reveals an emergence of auditory zones within each garden which can be

Figure 48: NURBS Map II (Kyu Furukawa Teien)

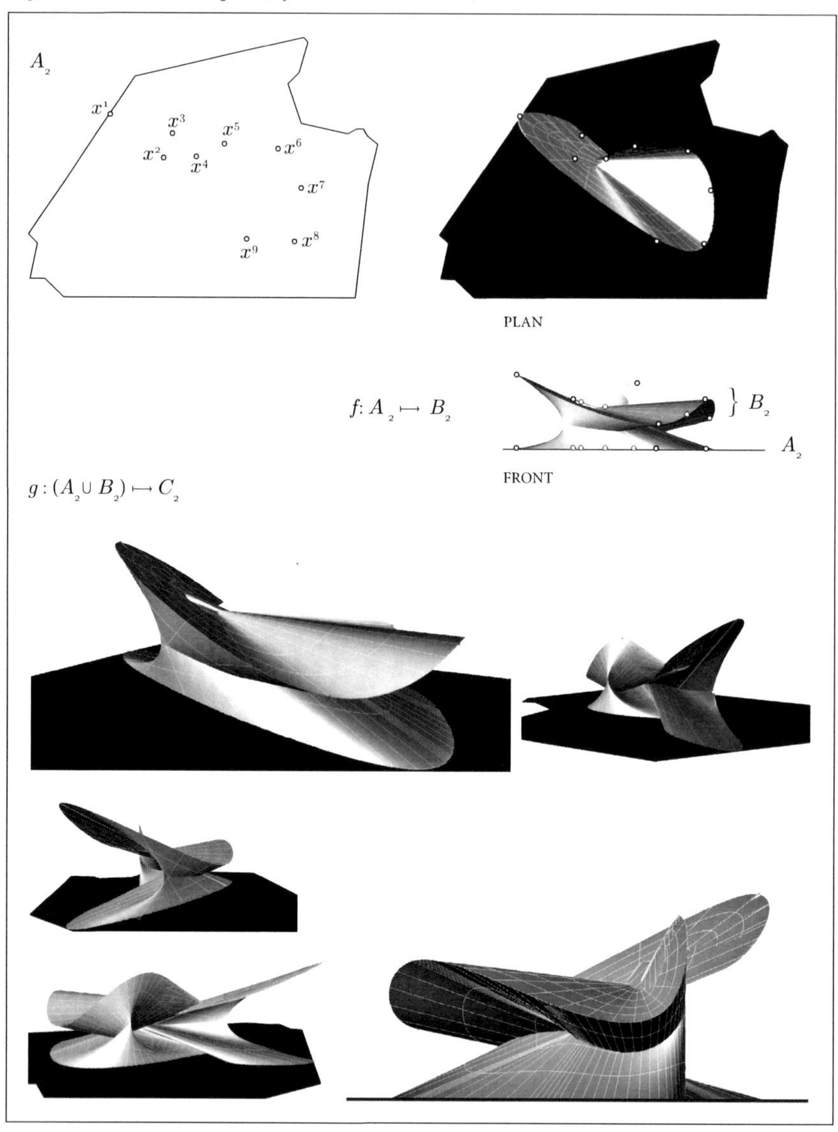

traced according to relative heights of the surface wall, volumetric distribution, manifold curvature and surface complexity regarding self intersections.

Figure 47 and 48 shows the extent to which the process of using lofting procedures sets the SPL_A data as a point elevation from the plan (0 dB). The process of lofting in the program Rhinoceros 3-D involves the generation of a surface from either a number of points or point-cloud or series of cross-section curves (which must either be closed or open, but not a combination of the two). A lofting algorithm generates a surface that connects each cross-section (points or curves) in a linear manner producing a planar topology that through lofting options in Rhinoceros 3-D can be closed (volumetric) or left open (manifold). Surfaces can be further refined through attributing various constraints to the surface and its "tightness" such as its tolerances regarding the control-points or control-curves.[49]

In each of the two closed-optioned forms generated from Koishikawa Korakuen (*NURBS Map I*) and Kyu Furukawa Teien (*NURBS Map II*), the theoretical 0dB A points are firstly located on a map of the garden and correspond to the sites that the SPL_A data was captured. These values are then located on the xy plane to which the collected SPL_A point values for B are projected onto the z-axis, thus forming a point array and series of guiding cross sections. But the location of the A points and the cross-referencing employed by the automated Rhinoceros 3-D lofting algorithm regarding the B points generates a series of inter-connecting surfaces within each 3-D map that are locally highly variant, and thus reflective of the range of the data set as a *continuum*.

The resultant 3-D maps can then be generalized in mathematical terms through the composite function:

$$g \circ f : A_n \longmapsto C_n$$

$$\textit{where } f : A_n \longmapsto B_n$$

$$g : (A_n \cup B_n) \longmapsto C_n$$

Here, the function g follows function f such that f represents the z-axis point elevation of SPL_A values and function g is the Rhinoceros 3-D lofting algorithm applied to all points A_n and B_n of each garden to produce surface C_n. The resulting composite function produces two unique 3-D objects, C_1 and C_2 that represent a materialization and flow of surfaces from the transmediation of $A_n \cup B_n$. The modeling then can be read as a transmediation of auditory information pertinent to each garden's auditory zones and the environmental invariants that construct the local soundscape qualities according to captured SPL_A values.

The transmediation reveals a particular set of congruencies. Of particular interest are the qualities of each *NURBS Map* (that is, surfaces C_1 and C_2) regarding the intersections, manifold complexity and converge in both maps around the entrance to each garden. Though each Map is fundamentally distinct in its general morphology, many localized topological features are evident. The most notable case is related to the data of the first 3 points $\{x^1, x^2, x^3\} \in B_1$, which drop in value by only half of that of $\{x^1, x^2, x^3\} \in B_2$, defining both sets and their subsets S_1 and S_2 gives:

49 Daniel Lordick, *Curve Surface Freeform: Rhinoceros for Architects* (New York: Springer-Verlag, 2009).

$$(n = 1) \rightarrow B_n = \{x^n, x^{n+1}, x^{n+2}, \ldots, x^{n+11}\},$$
$$S_n \subset B_n \mid S_n = \{(x^n, x^{n+1}, x^{n+2}) \mid n = 1, x^n \in B_n \wedge x^n \in \mathbb{R}\}.$$

$$(n = 2) \rightarrow B_n = \{x^n, x^{n+1}, x^{n+2}, \ldots, x^{n+7}\},$$
$$S_n \subset B_n \mid S_n = \{(x^{n-1}, x^n, x^{n+1}) \mid n = 2, x^n \in B_n \wedge x^n \in \mathbb{R}\},$$

$$\textit{and } (x^{n-1} - x^{n+1}) \in B_2 = 2(x^n - x^{n+2}) \in B_1.$$

This becomes a localized topological surface feature within each of the *NURBS Maps* around the garden entrances. Here the complexity of the surfaces is in contrast to the greater and more evenly distributed manifold to the rear or interior of the gardens. This is particularly evident in *NURBS Map I* (Figure 47), where a clear transition is present from the data that makes up the front of the form (the edge closest to the garden boundary: $\{x^1, x^2, x^3\} \in A_1$) to that of the quieter rear of the garden ($\{x^{10}, x^{11}, x^{12}\} \in A_1$). The smoother, less complex surfaces that exist at the rear of the object express this data trend. The corresponding SPL_A data from this zone showed a greater uniformity and had a smaller range of values. In comparison, the values from the garden's first 50-60m from the entrance show a greater variability, and are materialized as a less uniform surface on the model.

As an introductory exploration of NURBS modeling for spatial auditory data visualization, *NURBS Map I / II* has revealed that a qualitative approach to representing information from Koishikawa Korakuen and Kyu Furukawa Teien necessarily places the relationship between hard data and its representation as requiring an *intermediary* or agent. This is perhaps what Szalapaj was pointing to in his assessment of the relationship between CAD and designers:

"Design is a subject that requires not only the creation and development of design ideas, but also increasingly in contemporary architectural practice, the effective *expression* of these ideas within computing environments by *people*. The implications of CAD representations are extremely important in their relationship to design intentions, since the expression of design ideas in architectural practice is carried out and controlled by designers, and not by automated computer algorithms magically generating previously unimagined virtual architecture."[50]

The technique of lofting in *NURBS Map I / II* has been approached for its malleability and at times indeterminacy, and not simply as a way to completely abstract the information from site. Moreover, it has been utilized as a tool for re-imagining and repurposing the information as architectonic form. As a creative transmediation of the garden SPL_A data, architectonic mapping is perhaps not only a method for highlighting trends within the data not readily seen, but similarly as a means for approaching

50 Szalapaj, *Contemporary Architecture*, 4.

what Novak has proposed as a way in which to link the composition of architectonic form to a common root from the auditory realm. Spatial and architectonic visualization of auditory data goes beyond the concerns of a simple *translation* of SPL_A data into a highly accurate representation in 3-dimensions, but partakes more within the spirit of the gardener, sculptor and designer. What then might be a further excursion on the path towards architectonic composition using auditory information from a Japanese garden? What techniques and approaches might allow for the collection of spatial auditory information from the Japanese garden to become the simultaneous building blocks for sound design, musical composition *and* architectonic form finding?

Kyu Furukawa Teien Shūsaku

The number of extensive spatial sound recordings obtained from Kyu Furukawa Teien that formed part of the initial assessment of it spatial auditory traits and characteristics (as explained in chapter 3) were obtained using a recording array and multi-channel auralization environment that assessed the recordings in listening tests that mimicked the 4 and 5 channel site recording arrays (that is, in the lateral plane at -30°, 0°, 30°, 120°, 210°). Through this auralization and qualitative assessment of the soundscape qualities of Kyu Furukawa Teien, the concept for how the acoustic information from these recordings might be utilized within both a musical and architectural environment for composition arose. Though Novak's formation of a framework or synthetic environment for generating a real-time archimusic was not initially a primary goal of this project, at least a means for creating a meaningful framework as well as some initial experimentation with NURBS mapping as explored in *NURBS Map I / II* was viewed as a feasible option. In an effort then that aligns perhaps more to the concerns of Haruyuki Fujii, Kiyoshi Furukawa and Yasuhiro Kiyozumi and their spatial compositional study that explored "the things in common and [those] differences between the structure of the operations in architectural design and the structure of the operations in music composition,"[51] I devised a methodology that borrowed and expanded on some of some of the previous work I have already discussed. By essentially combining the approaches utilized in *Acoustic Intersections* as well as the data-structure visualization used in *Sesshutei as a Spatial Model*, a framework for utilizing spatial auditory information within Kyu Furukawa Teien was developed. The project, *Kyu Furukawa Shūsaku* (Kyu Furukawa Teien Study) takes as its focal point the large pond *shinjiike* and its immediate surrounds. The pond is such a central part of the garden and its auditory, material and timbre characteristics provide both important *soundmarks* as well as *passive aural embellishments* to the larger sound world of the garden. Thus a close analysis of the acoustic information and its potential to provide

51 Haruyuki Fujii, Kiyoshi Furukawa and Yasuhiro Kiyozumi, "Towards a Proposal of Method of Composing and Designing Architectural Integrating Music and Architectural Space," *IPSJ SIG Technical Reports* 19 (2006): 1-6.

transmediation strategies for musical composition and architectonic form generation was firstly undertaken via the programming framework of Pure Data.

Data-Structures

Building on the technique for musical parameter extraction that was first used in *Acoustic Intersections* through the Pure Data programming language and its native objects `fiddle~` and `bonk~`,[52] a five-channel recording from *shinjiike* was initially selected as the basis for both an architectonic mapping experiment and the representation of musical thematic materials for composing with pitch classes. As was similarly used in *Acoustic Intersections*, pertinent musical information such as cooked pitch, soundfile amplitude, raw attack spectrum and envelope attack was extracted via Pure Data's data-structures programming possibility. By generating a series of plain text files that recorded the unfolding of audio information from each channel of the recordings in real-time, a data bank of environmental auditory invariants was collected that reflected each channel's data-set range for:

- Raw attack spectrum in Hertz (an eleven element list recording values in eleven frequency bands).
- Soundfile amplitude in dB.
- Envelope attack recorded in seconds from soundfile onset.
- Cooked pitch in Hertz per 500ms.

To represent the data-sets in a qualitative way, while similarly allowing for a discrete interrogation of the data-set at any point in the soundfile's unfolding, meant that a visual representation (in the form of a quasi-score) needed to account for myriad parameter ranges, as well as various data extraction methods (Figure 49). Pd's data-structuring capabilities presented a viable method for representing the large amounts of information from the acoustic parameters described above. As Miller Puckette notes:

"Pd is designed to offer an extremely unstructured environment for describing data structures and their graphical appearance [. . .] To accomplish this Pd introduces a graphical data structure, somewhat like a data structure out of the C programming language, but with a facility for attaching shapes and colors to the data, so that the user can visualize and/or edit it. The data itself can be edited from scratch or can be imported from files, generated algorithmically, or derived from analyses of incoming sounds or other data streams."[53]

52 Miller Puckette, Theodore Apel and David Zicarelli, "Real-time Audio Analysis Tools for Pd and MaxMSP," in *International Computer Music Conference Proceedings* (Ann Arbor: ICMC, 1998), 109-112.

53 Miller Puckette, M. 2002. "Using Pd as a score language," in *International Computer*

Figure 49: Pd data-structure from recording at Kyu Furukawa Teien

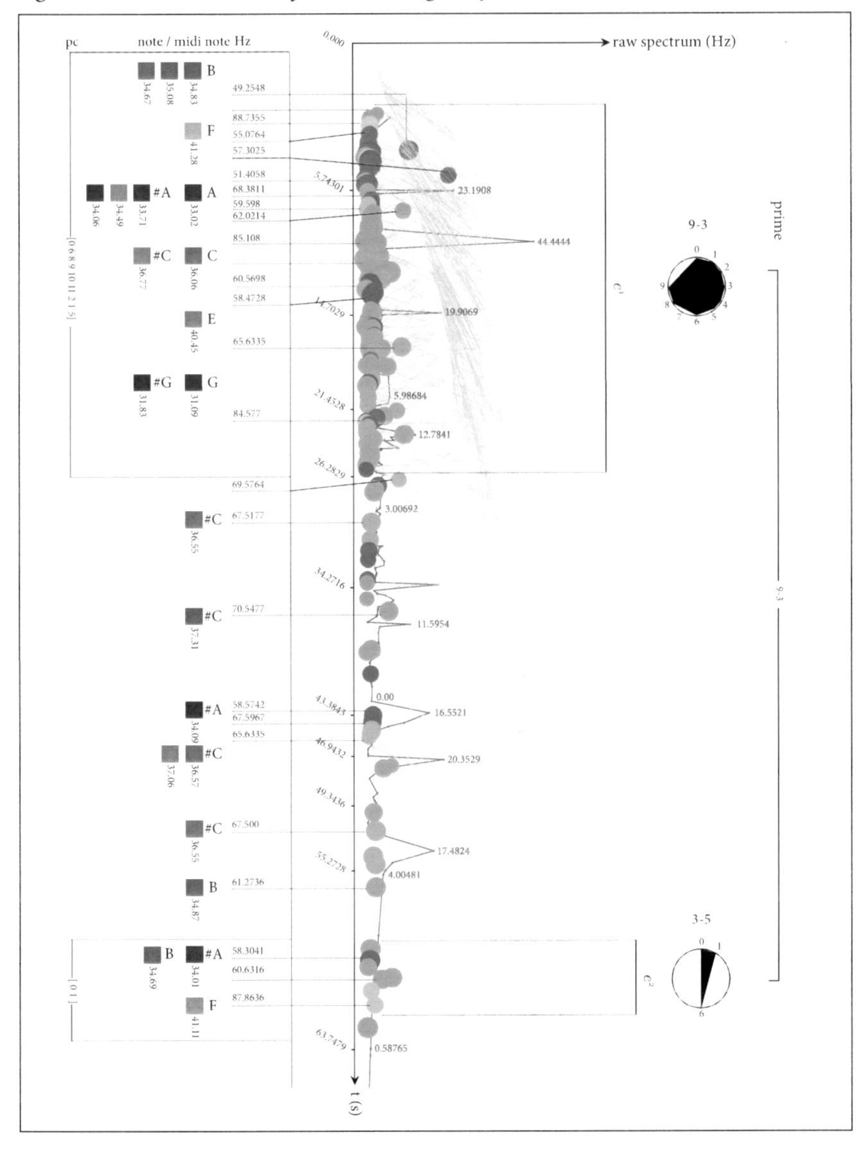

The developed Pd patch used in *Kyu Furukawa Shūsaku* then automated a real-time analysis and drawing process from the 5-channel soundfile through the following approach (square bracketed text is Pd programming language, "//" indicates explanatory comments):

Music Conference Proceedings (Gothenburg: ICMC, 2002), 187.

```
[struct v3-structure float x float x_l float y float y_l float z
float z_l float a float a_l]
```
//the basic patch structure and list of parameters (x, x_l, y, y_l, z, z_l, a, a_l) as float values//

```
[drawnumber 0 x y 100]
```
//number to draw (0), its position (x, y) and its color in RGB (100)//

```
[filledpolygon z y_l x_l 0 0 0 0 -0.2]
```
//fillcolor of the polygon (z), its outline-color (y_l), its line-width (x_l) and position (x, y), . . .//

```
[drawpolygon 777 1 x_l 0 y -10 y_l -10 a_l x y z 20 0]
```
//line-color (777), line-width (1), position (x, y), . . .//

//where x = raw spectrum of attack, x_l = soundfile amplitude in dB, y = envelope attack in ms, y_l = cooked pitch in Hz and z, z_l, a, a_l = 0 (undefined)//

The above data structure script of basic parameters was used to produce the central graphic element shown in Figure 49. The resulting data-structure mapping of channel 1 of the 5-channel soundscape recording of *shinjiike* produces an extended scroll-like representation that is read from top to bottom. The graphic score was produced in such a fashion as a means to approach what David A. Slawson notes as the qualities of a "scroll-garden," where the totality of a landscape scene is compressed and presented within a single frame:

"Now when a garden is viewed in scroll-garden fashion, all in a single frame, the shape of the site and the way it is framed—two factors that are intimately related—become much more important. Because so much attention is riveted on a single frame, [. . .] the artist—whether filmmaker, photographer or landscape garden designer—naturally lavishes great care on the way the composition in framed."[54]

For the Pd data-structure, as a function of time, this abstraction and compaction of the 'view' into a single digestible scene is mirrored in its unfolding from the top of the page downwards. Cooked pitch is tracked through the slight variations in color of the circles (via the `filledpolygon` command), with their relative size and situation accounting for changes in dB levels as a function of attacks detected in the envelope by `fiddle~`. The extensive line work of the grey polygons (`drawpolygon`) that spread outwards from the circles accounts for a combined mapping of all the acoustic parameters collected (by mapping points as polygon indices), while the `drawnumber` instruction inscribes a 0 at every occurrence of an envelope attack (y-axis) through

54 Slawson, *Secret Teachings*, 81.

bonk~, as a function of the first value in the eleven-frequency band list of raw spectrum attacks (x-axis).

In addition to the drawing inscription and mapping of raw acoustic information, the Pd data-structure has been annotated in an effort to extract pertinent and usable musical pitch space from the recording for sound design or composition. As is revealed in the annotated analysis of the data-structure, varying length macro-samples of audio data from channel 1 shows a range of pitch values evident at the microtonal level. The analysis includes a color legend in which slight variations in RGB (red, green, blue) value for envelope events (drawn as circles via the filledpolygon command) are mapped correspondingly as metrics in frequency (Hz), midi-note number (1-127), and musical note-name (from an equally tempered scale). The translation of the frequency to midi-note number and keyboard note-name was actuated through Pd's f2note object with 440Hz as the reference frequency. As can be ascertained from the data-structure, there is a large amount of pitch-timbre variation in the samples from *shinjiike*. Various RGB color combinations equate to a close proximity of frequency values to a central pitch indicating a host of microtonal clustering.

In an effort to further homogenize the pitch data so that these localized neighborhood relationships could be formed into larger related pitch groups, a pitch-class set theory framework was utilized so that the musical pitches found via the data-structure could be generalized and represented in absolute form regardless of their location in the relative octave. Allen Forte's pitch-class set theory[55] proposes that the eleven semitones of the equally tempered octave, {C, C#, D, . . . B}, are mapped to a set of integers such that $\mathfrak{T} = \{0, 1, 2, \dots 11\}$. The set $\mathfrak{T}$ is a subset of the set of natural numbers, $\mathfrak{T} \subset \mathbb{N}^0$, for which the elements of $\mathfrak{T}$ obey the congruent modulo n relation such that $a \equiv b \pmod{n}$. Here, for example, the musical pitch B can be understood as both 11 (eleven semitones upwards from C) or -1 (one semitone downwards from C), and given that $a - b$ is an integer that is a multiple of n, Forte's pitch-class set theory is said to be conceived in modulus 12.

This generalization of musical pitches enables new compositional approaches to emerge that are related to the twelve-tone or tone-row compositional method of Arnold Schoenberg, Anton Webern and Alban Berg of the second Viennese School, and the later extension of the method by Karlheinz Stockhausen, Milton Babbitt and Pierre Boulez.[56] Here, musical space is homogenized such that octave occurrence, or the relative location of the musical pitch is of less importance than its quality (that is, its note name). For example, C_4 (or middle C) is considered structurally the same as C_7 (the highest C on a keyboard) within a compositional context. But Forte's theory not only accounts for a less hierarchical approach to constructing musical pitch space, but also for the ordering and classification of grouped subsets of the eleven pitch classes.

55 Allen Forte, *The structure of atonal music* (New Haven, CT: Yale University Press, 1973).

56 J. Grant Morag, *Serial music, serial aesthetics: composition theory in post-war Europe* (New York: Cambridge University Press, 2001).

Indeed Forte was the first to name all possible combinations of three to eleven element subsets using the naming convention of *c-d*, where *c* indicates the cardinality of the set and *d* is the ordinal number.[57] For example, the three-element subset or trichord [A, A#, B] is comprised of the elements {0, 1, 2} where in its prime form (that is, its most compact form) is comprised of C = 0, C# = 1, D = 2 and is labeled by Forte as 3-1: that is, the first ordered subset of cardinality 3. Thus the trichord 3-2 = {0, 1, 3}, 3-3 = {0, 1, 4}, etc.

In the *shinjiike* data-structure, natural or equal tempered pitch-class formations were obtained through accounting for larger phrase groups in the samples by defining pitch-class sets according to the appearance of obvious formations. For example, where there is a clustering of circles in the site sample, a continuous unfolding and overlapping of data causes a pattern of coagulation. Larger groups or umbrella collections, E^n were found by observing constituent elements and subsets (see groupings marked e^n for the subset partitioning), in which discrete Forte trichords dictated the articulation of the data on its most elemental harmonic level. For the data-structure of *shinjiike* two prominent Forte subsets emerged, 3-5 as well as an appearance of 3-2 within the larger collection belonging to 9-3:

$$
\begin{aligned}
E^1 &= \text{9-3}\,\{0, 1, 2, 3, 4, 5, 6, 7, 8, 9\},\\
\textit{where}\ e^1 \in E^1 &= \text{3-2}\,\{0, 1, 3\},\\
e^2 \in E^1 &= \text{3-5}\,\{0, 1, 6\}.
\end{aligned}
$$

In addition to the data-structure developed for the recording at *shinjiike*, two other sites around *shinjiike* were mapped, namely *ootaki* and *karetaki*. Table 9 shows the Forte pitch-class collections obtained from these sites including the macro collection E^n and subgroup e^n.

Table 9: Pitch-class collections and subsets from Kyu Furukawa Teien

Site name	E^n	$e^n \in E^n$
ootaki	[8-5] for $n = 1$	[3-1] [7-z38]
shinjiike	[9-3] for $n = 2$	[3-2] [3-5]
karetaki	[8-9] for $n = 3$	[4-5] [5-20] [3-4] [3-9] [3-5]

57 Morag, *Serial music*, 12.

This method of equal tempered pitch information extraction facilitates both a generalized proportioning of the subgroup constituents, e^n, as well as providing signature pitch-class collections for the site sample, E^n. Amplitude variations in the sample data could also be examined through observations in the deviation of the line connecting the `drawnumber` values (marked with 0). Here, the values indicate the relative loudness of the envelope attack (from the `bonk~` object) for the first of an eleven-frequency band list. The incoming signal is measured as a relative strength across bands with center frequencies of 100, 300 and 500 Hz (bandwidth 200 Hz), with the remaining eight tuned to each half-octave above 500 Hz so that the top one is centered at 8 kHz. Any of the other values within the eleven bands for each site sample can be called up and used to construct a similar template for constructing a dynamics chart, or transformational network.

Form Generation

That there were 3 distinct pitch-class sets, E^1 [8-5] at *ootaki*, E^2 [9-3] at *shinjiike* and E^3 [8-9] at *karetaki* found in the recordings taken from the various sites also presented a case for the use of this information in developing an architectonic study or *shūsaku*. This process is documented in Figure 50. In an effort to incorporate spatial elements inherent in the site, as well as musically derived data from the soundscape, *shinjiike* was conceptualized as a *pitch-class pond* that fed the three samples sites with data. Using Rhinoceros 3-D, seventy-seven control-points ($cp_i \in \mathfrak{P}$) populated the confines of the general area of the pond as a point-cloud and were randomly elevated through eleven different heights that corresponded to the range of Forte's modulo twelve pitch-class series classification system: this means that the values of individual pitch-classes recorded at each site can be spatially articulated.

According to the pitch-class constituents inherent in each site's defining series means that various trajectories can be composed to traverse the pitch-pond and inscribe these values with NURBS polylines. When a possible closed trajectory is defined, it can be used as the cross-section curve of a closed loft, thus enabling the creation of a surface as previously explored in *NURBS Map I / II*. A vast number of trajectories may be nominated for defining the pitch-class values for each sample site's pitch-class series. This particular aspect of *Kyu Furukawa Teien Shūsaku* is a natural usurpation of Cagean aesthetics regarding indeterminacy as a process to drive diversity, experimentation and the exploration of new territories. When indeterminacy is relevant to the composition of a work, then the designer or composer becomes the agent to which the delivery of the work is necessary. In the case presented here, a particular collection of realizations of an indeterminate process becomes a representative of one group-version of outcomes. By presenting three different architectonic versions of the same indeterminate process, an understanding of the nature of indeterminacy as a pointer to an underlying larger structure is established.

Figure 50: Shūsaku form generation process

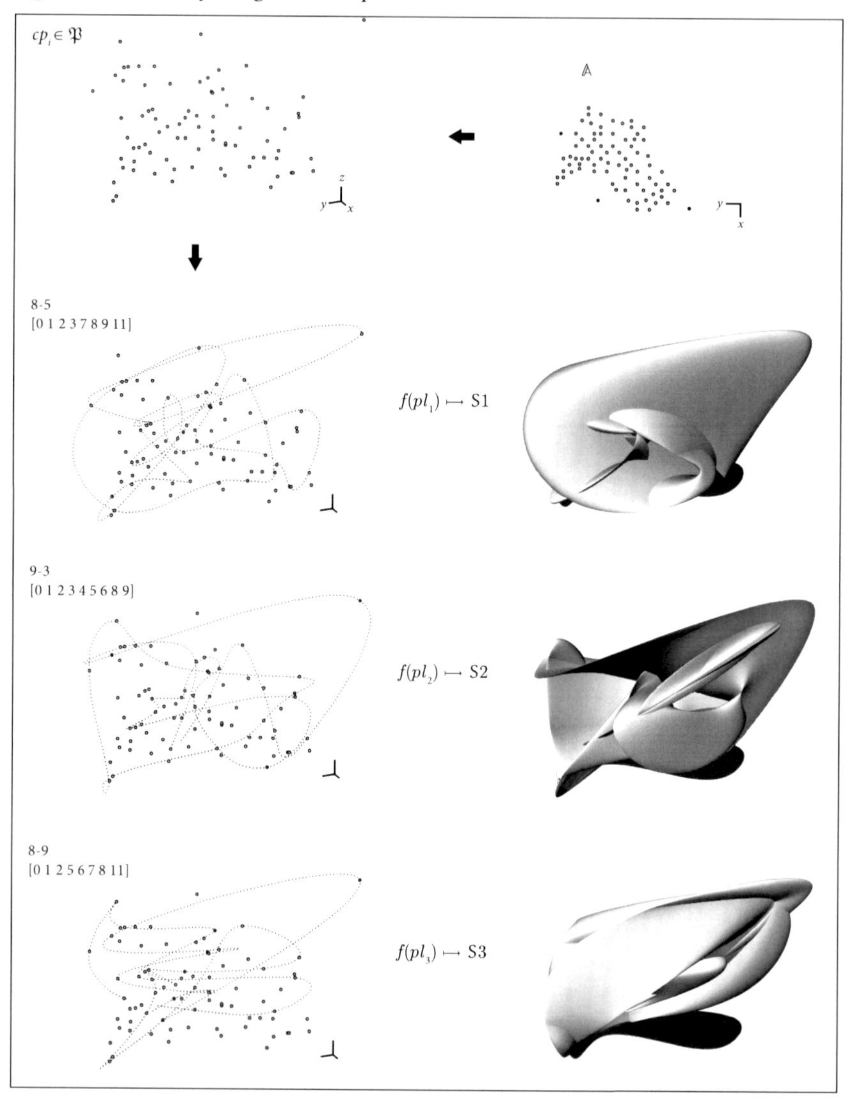

Indeed, modern computer processing capabilities can be a valuable tool in the production and performance of indeterminate systems for spatial composition. Given the real-time capabilities of a program like Pd and Rhinoceros 3-D, and the nature of data manipulation and distribution within the programs, an indeterminate system explored through an infinite number of variations provides that system with self-referential meaning and a particular identity because of its ability to explore every possible face of an indeterminate process.

Figure 51: Shūsaku architectonic NURBS forms

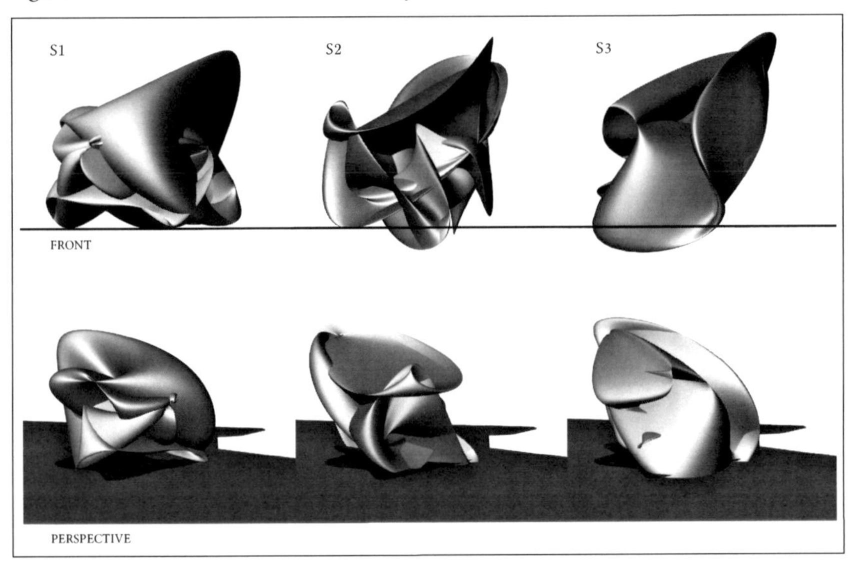

Figure 51 documents three variations in architectonic form—S1, S2, S3—that are mappings from three possible trajectories of closed polylines for each sample site's pitch-class collection. Here, the surface complexities of the three architectonic forms are naturally highly divergent given the randomness of the trajectory paths and the potential for indeterminate process to guide the decision-making. Each surface, Sn, is produced via the Rhinoceros 3-D lofting function f, from a closed NURBS polyline, pl_n such that the z-axis values for the control-points $cp_i \in \mathfrak{P}$ are also elements of $\mathfrak{T}$, for which the (x, y) values are randomly located (via function r) within the area $\mathbb{A}$ which is the boundary of shinjiike and a subset of the garden $\mathbb{G}$:

$$f(pl_n) \longmapsto Sn$$
$$\big((z \in cp_i) \in \mathfrak{T} \wedge \mathfrak{P} \mid \forall cp_i \in pl_n\big) \rightarrow \big(\mathrm{r}(x, y) \in cp_i \subset \mathbb{A}\big) \subset \mathbb{G}$$

Though the points and their values remain fixed between the surface generations of Sn, the interaction (crossings, shared points etc.) of the NURBS polylines through the pitch-class pond causes localized permutations, interior folds and self-intersections between the forms of Sn.

The transmediation of the Forte pitch-class collections then goes somewhat beyond the strategies of SPL_A mapping in *NURBS Map I / II* and more into the realm of utilizing spatial information both for musical ends as well as for architectonic form generation. But any consideration of the 3-D forms of *Kyu Furukawa Teien Shūsaku* as an invitation for primarily producing architectonic gesture also means that a conventional architectural program is to be subverted. For this investigation, such intentions are resolved by the manner in which the original spatial information is transmediated into three-dimensions. A surrogate program does establish itself when one considers that the meaning of the form comes from the construction process *itself*, and the ways in which the site of the garden and its place are a function of this manufacturing. As a spatial concept then, such a realization sits *before* the standardized architectural object or architectural model: it is a proto-architecture in that it represents the conditions under which architecture might proceed, and under which particular spatial programs might develop. This is an equal approach as to the other half of the investigation where the extraction of Forte pitch-class collections for musical compositional systems or sound design strategies is presented as a *tool* for composition to unfold. In a similar manner to the end results of *Re-Ryoanji*, the pitch-class extraction methods used in *Kyu Furukawa Teien Shūsaku* are intended as an available compositional strategy. Rather than providing a specific usurpation of the pitch collections within a derived sound design, the aim of the project has been to investigate the point of common connection between architectonic form and musical form building regarding the Japanese garden as a spatial paradigm. The results of *Kyu Furukawa Teien Shūsaku* do then attest to the exploration of an area that Novak has championed but rather than containing this nexus within the confines of a synthetic environment or a "dataworld," the point of common contact between the compositional methods before they diverge into the contexts of the musical or the architectonic has been purely via a creative usurpation of spatial information.

Indeed, in a position that diverges from Novak's, architect Steven Holl's positioning of architecture and music as potential interdisciplinary expressions is argued through the uptake of the concepts of "number, rhythm, notation and proportion"[58] as spatial concepts for both disciplines. For Holl, this approach can only be realized through the exploration of a new territory in which music and architecture become *undefined parameters* and act as catalysts for a new spatiality. In such a paradigm a shared vocabulary and domain of concepts and objects would allow for constructions of form in both tangible and intangible materials as well as a new intimacy to rise between architecture-space and music-space. Perhaps by reading *Kyu Furukawa Teien Shūsaku* as a window into such a domain confers its trans-disciplinary role for generating such an ontology. The example of *Kyu Furukawa Teien Shūsaku* thus also explores what Yeoryia Manolopoulou sees as the "impulsive or systematic"[59] processes that arise within an aleatoric design process, and thus perhaps offers subjective value to the perception of the indeterminate process as design-destructive. But the architectonic forms and pitch-class schema of *Kyu Furukawa Teien Shūsaku* also occupy an analogous territory to the ways in which Robyn Evans see architectural drawings as

58 Steven Holl, "Stretto House," in *Architecture as a translation of music* (New York: Princeton Architectural Press, 1994), 56.
59 Yeoryia Manolopoulou, "The Active Voice of Architecture: An Introduction to the Idea of Chance," *Field* 1 (2007): 70.

sitting prior to reality.[60] They are spatial potentials that share a lineage to all other types of material (tangible or intangible) constructions or models born from the same process.

Perhaps the most pertinent aspect of the project work of *Kyu Furukawa Teien Shūsaku* in regards to the ongoing investigation into transmediation techniques for utilizing spatial information within the Japanese garden has been around the issues concerning appropriate methodologies in which to traverse and bridge the domains of spatial analysis and spatial composition. Architectonic form generation here has been only a suggestive measure for enabling a conceptual series of architectonic sketches to be created. What might a more concrete, program specific approach to bridging this divide yield? In the final project of this chapter I introduce and document the project *«No-Stop Garden»*. Here, the approach of generating spatial information from an analysis of Kyu Furukawa Teien is again undertaken, for which the form finding strategy uses mathematical strategies and rule-based approaches as the premise for creating a series of virtual landscapes via the parametric coding environment Grasshopper 3-D. *«No-Stop Garden»* thus brings the entire investigation on transmediation methodologies back home to Japanese landscape gardening by examining how the garden's fundamental spatial qualities, as mathematical principles, may guide a generation of imagined future landscapes.

«No-Stop Garden»

Perhaps one of the most iconic images that emerges from the work of the 1960s Italian design studio Archizoom Associati (that comprised the team of Andrea Branzi, Giberto Corretti, Paolo Deganello, Massimo Morozzi, Dario and Lucia Bartolini) involves their conceptual models within the 1968 project *«No-Stop City»*.[61] One particularly striking image of the project acts out a particularly radical and somewhat dystopian view of the future city as a perfect, infinite, spatial repetition. A small completely flat base houses a perfect single white rectangular prism: a building, windowless and seemingly impenetrable though highly refined. The building's flat green roof is adorned with a decorative white pattern suggestive of a series of garden paths interspersed with straight lines and other geometric shapes. Within the clinically measured expanse and exactness of the ground base are a few trees (again set out in perfect rows) and hedges as well as a small collection of stylized high gabled forms that suggest those typical houses of suburbia as well as a collection of other smaller scale cubic prisms—all uniformly white in color. Straight paths or access roads feed both these small dwellings as well as the larger feature building located at the center. The use of mirrors on all four walls manages to take this spatial paradigm into the infinite through a series

60 Robin Evans, *Translations from Drawing to Building and Other Essays*, (London: Architecture Association, 1997), 186.
61 Andrea Branzi, *No-Stop City: Archizoom Associati* (Orleans: Hyx, 2006).

of endless repetitions seen through every 'wall' of the enclosure via compound reflections. Spreading out into infinity, the repeated forms approach a delirious conformity in which individual expression is a trait that can only become seemingly expressed *within* the windowless forms. The project then suggests that the habitat of the central prism and its interior qualities are the only means in which to break the spatial repetition. As Jane Alison suggests, Archizoom considered architecture as:

"an intermediate category of urban organization that had to be surpassed, No-Stop City makes a direct link between the metropolis and furniture: the city becomes a succession of beds, tables, chairs and wardrobes, domestic and urban furniture meld into one. We respond to qualitative utopias with the only possible utopia: that of Quantity."[62]

Archizoom use the model as a means to communicate how urban infrastructures can become devices for enacting control, conformity or disorientation on the experience of the urban environment, and similarly, how the urban condition itself is an inevitable function of mass production. This particular architectural model has played an important role for my project «*No-Stop Garden*» whose generative process, philosophy and delivery acts as somewhat of a counter to the dystopian discourses created by Archizoom's iconic design thinking, while seeking to continue and extend the transmediation strategies of the usurpation of the Japanese garden as a sign system.

«*No-Stop Garden*» then can be regarded as an experimental design methodology that seeks to usurp spatial qualities of the garden Kyu Furukawa Teien as parameters for generating a vast continuum of possible future landscape designs. Rather than projecting, as Archizoom do, an infinitely repeating singular architectonic form, «*No-Stop Garden*» instead seeks to explore the limits of architectonic morphology through automated variation. One of the ways that this is explored is by playing off Robert Woodburry and Andrew Burrow's notion of *design space exploration*. Here, design space represents both a conceptual and actual space in which myriad possibilities for spatial form are mapped and explored such that a completed or resolved end point or end design is considered not a singular entity or as an absolute, but rather a function of the iterative process of *searching*. The authors explain that

"design space exploration is the idea that computers can usefully depict design as the act of exploring alternatives. This involves representing many *designs*, arraying these represented designs in a network structure termed the *design space*, and exploring this space by traversing paths in the network to visit both previously represented designs and to find sites for new insertions into the network."[63]

62 Jane Alison, Marie-Ange Brayer and Frédéric Migayrou eds., *Future City* (London: Thames and Hudson, 2006), 276.

63 Andrew Burrow and Robert F. Woodbury, "Whither Design Space?," *Artificial Intelligence for Engineering Design, Analysis and Manufacturing* 20 (2006): 63–82.

This is quite a powerful idea for design praxis, particularly given the recent rise in CAD, and moreover, the recent emergence of parametric modeling environments.[64] I have quite directly incorporated the notion of design space within *«No-Stop Garden»* not simply as in the guise of Archizoom's act of repetition as an *ad nauseam* driver, but more subtly, as a means to produce *design abundance* under particular constraints. These constraints, or spatial dicta in *«No-Stop Garden»* arise from the wealth of analysis already outlined within this book regarding the Japanese garden. Thus the fundamental premise of *«No-Stop Garden»* is to read the Japanese garden as a spatial database, by which a transmediation of its proclivities will give rise to outcomes that are in themselves explorations of its design space. *«No-Stop Garden»* then positions the Japanese garden as a spatial exemplar, and then interrogates this paradigm by utilizing an algorithmic approach to creating new landscapes that are at once explorations of the pure possibilities of design space navigation as a function of parametric modeling, while at the same time re-appropriations of *deep-level* spatial attributes of Japanese garden design.

Indeed, the rapid increase in digital tools for design, and within the last five years, parametric approaches to spatial design in architecture have shifted the traditional notion about the fixidity or resoluteness of geometry and the completeness or desired longevity of architectonic form. Architect Mark Gulthorpe notes these new opportunities have created a situation in which a design and the design process has moved away from the construction of a timeless monument and towards the construction of a *continuum* in which myriad possibilities can be explored intuitively in real-time. He notes that "parametric modeling offers the possibility to build variance into a system, to model relations, such that shifting the parameters creates variability."[65] Indeed, the basic notion regarding parametric modeling in digital architectural environments is that those fundamental elements of modeling—points, surfaces, curves and lines—can be defined in relation to each other or to other geometries or metrics (such as algebraic rules). Thus chains of relations can be defined, for which the relations can be fixed (in a linear or non-linear manner) or themselves fluid (that is, controlled in real-time through GUI user input) such that when one of the elements—acting as a parameter—is manipulated, moved or projected, the rest of the chain responds in a fashion that preserves the defined parameter relations. Design then becomes algorithmic in that the development of form is achieved through the careful crafting and consideration of the underlying structures, framework, or relationships between elements rather than the traditional attention to specific elements *in isolation*. Algorithmic design then is concerned with the system as a totality, in which every part of the algorithm is adjustable, yet capable of influencing the rest of the structure. Patrik Schumacher describes this new paradigm in much the same way that Novak conceived of the liquid nature of the "dataworld":

---

64 Cristoph Gengnagel, et al. eds., *Computational Design Modeling - Proceedings of Design Modeling Symposium Berlin* (Berlin: Springer-Verlag, 2011).
65 Mark Gulthorpe, *The Possibility of (an) Architecture* (New York: Routledge, 2008), 135.

"Instead of assembling rigid and hermetic geometric figures–like all previous architectural styles—parametricism brings malleable components into a dynamical play of mutual responsiveness as well of contextual adaptation".[66]

If Archizoom's *«No-Stop City»* then is an exploration of a vast and dystopian urban future enacted through continuous repetition in which the fixed nature of the architectonic serves as the underlying conceptual framework, *«No-Stop Garden»* embodies an antidote where a variable utopian landscape form is continually generated via an embrace of changeability and fluidity. This approach is perhaps both an acknowledgement of the power of Cage's embrace of the experimental or aversion to predetermination or premeditated compositional methods as much as an embrace of what Gustaf Almenberg describes as Umberto Eco's concept of the "open work" and the traces of the nineteenth century Symbolist movement. Almenberg notes that the "characteristic for [Eco's] Open Work is that it remains *unfinished* without an intervention of some kind from the spectator / listener / reader / performer / consumer."[67] *«No-Stop Garden»*, in a similar manner, seeks an unending variability in which rather than relying on the continued re-use of a monumental geometry that seeks to sit outside of time, uses time as the basis for which a continued evolution drives a utopian variability.

Kyu Furukawa Teien Lattice as Piecewise Function

I explored in chapter 3 the use of FCA as a tool that revealed some of the connections and disconnections occurring between design concepts as they pertain to landscape forms and sound forms within Kyu Furukawa Teien. The methodology of FCA is one primarily concerned with generating a taxonomy to describe a closed universe of objects and their attributes. The conditions by which relations are formed within the visualized lattice, $\underline{\mathfrak{B}}(\mathbb{K})$, gives rise to the notion that these relations have intrinsic mathematical characteristics, and therefore, can be said to *describe* the garden in absolute and fundamental terms. But if they exist in abstract mathematical terms, they might then be useful not only in terms of describing these extent relations, but more interestingly, be used to *prescribe* relations within a new sign system. This is perhaps at the heart of Siegel's notion of transmediation, though the manner in which the signs are not only firstly identified, but then usurped and re-constructed within a new sign system, can be approached in myriad ways.

66 Patrik Schumacher, "Parametricism and the Autopoiesis of Architecture," *Log* 21 (2011): 63-79.
67 Gustaf Almenberg, *Notes on Participatory Art: Towards a Manifesto Differentiating it from Open Work, Interactive Art, and Relational Art* (Central Milton Keynes: AuthorHouse, 2010), 91.

The underlying premise of «*No-Stop Garden*» then has been to develop a method by which the Hasse diagram, $\mathfrak{B}(\mathbb{K})$, of Kyu Furukawa Teien ceases to be an analytical tool and begins life anew as a catalyst for informing a design method to emerge. But to achieve this goal for inverting $\mathfrak{B}(\mathbb{K})$ from a descriptive diagram into a prescriptive one, I envisaged reading the lattice as a type of mathematical function in and of itself. By imagining the partially-ordered lattice (see Figure 26) as a function whose set of input variables, that is, the set of all attributes M in $\mathfrak{B}(\mathbb{K})$ map to the set of all objects G via the *reading rule*, enables a new paradigm to develop that can better aid an approach to designing new parametrically controlled landscapes. As a first step, the seven attributes AAE (active aural embellishment), PAE (passive aural embellishment), Ter. (terrestrial), Nat. (natural), Sea.(seasonal), Arch. (architectonic), and Sou. (soundscape) are envisaged to be randomly seeded with values (a) such that the domain of the function can be written as:

$$Dom\,(f) = \{a \in \mathbb{R} \mid \forall a : 0 < a \leq 10\}.$$

The domain, or range of outputs of the function, has intentionally been limited to real numbers greater than zero but less than or equal to 10 in an attempt to both simplify the mechanism for computing the values of M, as well as to provide a more workable range for translating into landscape features, or enumerating landscape elements within Rhinoceros 3D and its parametric plug-in software Grasshopper 3D. The structure then of $\mathfrak{B}(\mathbb{K})$ will produce values for the objects that are located at the bottom of the lattice. These values can then be used to guide design decisions or inform scripting routines for generating landscape features. But this method also requires that the nodes $\{\mathfrak{c}_\alpha, \mathfrak{c}_\beta, \mathfrak{c}_\gamma\}$ and $\{\mathfrak{c}_\delta, \mathfrak{c}_\varepsilon, \mathfrak{c}_\zeta\}$, of the lattice come into play as points where data streams (i.e., values of a) must be transformed in some fashion. I use these six nodes in the simplest fashion, such that:

$$\begin{aligned} &f(x, y) = x + y \\ &\quad \textit{where } x = a \in \mathrm{int}(\mathfrak{c}_x) \\ &\qquad\qquad y = a \in \mathrm{int}(\mathfrak{c}_y) \end{aligned}$$

Here, the values of a in the intention (upward edges x and y) of the node $\mathfrak{c}_n$ are under consideration such that those values are summed at the node. This simple additive solution then utilizes the universal property within $\mathfrak{B}(\mathbb{K})$ that:

$$\forall \mathfrak{c}_n \in \left(\bigcup \mathfrak{E} \cup \bigcup \mathfrak{D}\right) \rightarrow |\mathrm{int}(\mathfrak{c}_n)| = 2$$

That is, every node ($\mathfrak{c}_n$) within the sub-lattices of $\bigcup \mathfrak{E}$ and $\bigcup \mathfrak{D}$ contain exactly two edges (or data streams) that converge. But because of the interrelation between attributes and their extension to multiple objects, variations in value of a among the set of attributes will necessarily effect a number of objects and not only a single object. This parametric aspect is a favorable condition for design given that some unknowns are likely to emerge regarding values computed for the objects. But to enable an easier approach to integrating the two sets of attributes M, and objects G, these sets are firstly renamed:

$$\textit{Let}\ \mathfrak{A} := \{\text{AAE, PAE, Ter., Nat., Sea., Arch., Sou.}\}$$

$$\textit{Let}\ \mathfrak{O} := \left\{\begin{array}{l}\text{fauna, exterior sounds, moving water, flora, earth,}\\ \text{still water, rocks, topography, paths, habitable architecture}\\ \text{bridges, garden ornaments}\end{array}\right\}$$

Each of the 7 formal attributes (without extensions) of $\underline{\mathfrak{B}}(\mathbb{K})$ then become the set $\mathfrak{A}$, with the 8 objects of $\underline{\mathfrak{B}}(\mathbb{K})$, the set $\mathfrak{O}$. But each of the formal attributes (Ter., Nat., Sea., Arch., Sou.) plus the paired attributes (AAE, PAE) existing on the same node can also be considered as sets in themselves for which the subsets of $\mathfrak{A}$ (indexed as A_1,… A_7), when in union are defined as:

$$\bigcup_{i=1}^{7} A_i = \mathfrak{A} := \{a \in \mathbb{R} \mid \forall a \in A_i, 0 < a \leq 10\}.$$

$$\textit{where}\ A_1 = \{\text{Sou.-AAE}\},\ A_2 = \{\text{Sea.}\},\ A_3 = \{\text{Nat.}\},$$
$$A_4 = \{\text{Art.}\},\ A_5 = \{\text{Ter.}\},\ A_6 = \{\text{PAE}\},\ A_7 = \{\text{Arch.}\}.$$

Here, the paired attributes of AAE and Sou. are combined so as to become a singleton of the set A_1, and thus are only assigned one value of a. Essentially «*No-Stop Garden*» has evolved then from the premise that after a random seeding of $a \in A_i$, where $\{a \in \mathbb{R} : 0 < a \leq 10\}$, values that define the objects can be found through utilizing $f(x, y) = x + y$ at those nodes $\mathfrak{c}_n \in (\bigcup \mathfrak{E} \cup \bigcup \mathfrak{D})$. This approach then enables a new framework for utilizing the structure of $\underline{\mathfrak{B}}(\mathbb{K})$. The parametric qualities of the lattice (regarding attribute extensions to multiple objects) means that after a random seeding, successive generations of object values can be explored, in effect creating a method by which the design space (i.e., the myriad possibilities of unique object counts or object combinations) produced by $\underline{\mathfrak{B}}(\mathbb{K})$ can be implicated as a function of time. In approaching the entire lattice then as a framework for the enumeration of objects and the application of a metric that connects attribute values directly to an object count, such as under the general conditions of $f(a) = \sigma_n$ (where $\sigma_n \in \mathfrak{O}$), means that a single comprehensive function that describes this framework is a necessity for design.

Below is the piecewise function I developed that describes $\mathfrak{B}(\mathbb{K})$ as a numerator for generating values for each of the objects according to lattice's node structure:

$$f(a) = \begin{cases} \sigma_1, & \text{if } a = \sum_{i=1}^{3} A_1 \\ \sigma_2, & \text{if } a = \sum_{i=1}^{3} A_{2^{i-1}} \\ \sigma_3, & \text{if } a = \sum_{i=1}^{3} A_{2i-1} \\ \sigma_4, & \text{if } a = \sum_{i=2}^{4} A_{i+\lfloor\sqrt{2i-1}\rfloor} \\ (\sigma_5, \sigma_6, \sigma_7), & \text{if } a = \sum_{i=3}^{3} A_{i+\lfloor\sqrt{2i-1}\rfloor} \\ \sigma_8, & \text{if } a = \sum_{i=1}^{4} A_{i+2} \\ \sigma_8, & \text{if } a = \sum_{i=1}^{3} A_{i+3} \\ (\sigma_{10}, \sigma_{11}, \sigma_{12}), & \text{if } a = \sum_{i=1}^{3} A_{i+4} \end{cases}$$

where $\sigma_1 =$ fauna, $\sigma_2 =$ exterior sounds, $\sigma_3 =$ moving water, $\sigma_4 =$ flora,
$\sigma_5 =$ earth, $\sigma_6 =$ still water, $\sigma_7 =$ rocks, $\sigma_1 =$ topography, $\sigma_1 =$ paths,
$\sigma_{10} =$ habitable architecture, $\sigma_{11} =$ bridges, $\sigma_{12} =$ garden ornaments

Values for $\sigma_n \in \mathfrak{O}$ are thus computed through various sums of values of $a \in A_i$. But by considering the set of formal objects $\mathfrak{O}$ of $\mathfrak{B}(\mathbb{K})$, as a union of eight sets, particular properties arise in the indexed subsets. The subsets in union are defined as:

$$\bigcup_{i=1}^{8} E_i = \mathfrak{O} : \{\sigma_n \in \mathbb{R} \mid \forall \sigma_n \in E_i, 0 < \sigma_n \leq 40\}.$$

$$\text{where } E_1 = \{\sigma_1\}, E_2 = \{\sigma_2\}, E_3 = \{\sigma_3\}, E_4 = \{\sigma_4\},$$
$$E_5 = \{\sigma_5, \sigma_6, \sigma_7\}, E_6 = \{\sigma_8\}, E_7 = \{\sigma_9\}, E_8 = \{\sigma_{10}, \sigma_{11}, \sigma_{12}\}.$$

It is perhaps pertinent to note here that there are two special cases that arise within $\mathfrak{O}$, in particular, the subsets E_5 and E_8, where because of the nature of $\mathfrak{B}(\mathbb{K})$ to contain object nodes with more than one object, I have assigned an equivalence between values of $\sigma_n \in E_5 \wedge \sigma_n \in E_8$ such that:

$$f(a) = (\sigma_5, \sigma_6, \sigma_7) \text{ if } a = \sum_{i=3}^{3} A_{i+\lfloor\sqrt{2i-1}\rfloor}$$

$$E_5 = \{(\sigma_5, \sigma_6, \sigma_7) | \sigma_5, \sigma_6, \sigma_7 \in \mathbb{R}, (\sigma_5 \sim \sigma_6 \sim \sigma_7) > 0 \wedge \sigma_5 + \sigma_6 + \sigma_7 \leq 30\}.$$

$$f(a) = (\sigma_{10}, \sigma_{11}, \sigma_{12}) \text{ if } a = \sum_{i=1}^{3} A_{i+4}$$

$$E_5 = \{(\sigma_{10}, \sigma_{11}, \sigma_{12}) | \sigma_{10}, \sigma_{11}, \sigma_{12} \in \mathbb{R}, (\sigma_{10} \sim \sigma_{11} \sim \sigma_{12}) > 0 \wedge \sigma_{10} + \sigma_{11} + \sigma_{12} \leq 30\}.$$

Another feature of $f(a) = \sigma_n$ is in the case of E_6 where the output range for values $(0 < \sigma_8 \leq 40)$ is larger than other cases in E_i and arises from the intention of the object 'topography' which includes the four attributes {PAE, Ter., Art., Nat.}:

$$f(a) = \sigma_8 \text{ if } a = \sum_{i=1}^{4} A_{i+2}$$

$$E_6 = \{\sigma_8 | \sigma_8 \in \mathbb{R}, 0 < \sigma_8 \leq 40\}.$$

The relationship then between the inputs and outputs of $\mathfrak{B}(\mathbb{K})$, in this guise, becomes much more tangible as a framework for the generation of a formal landscape structure that behaves parametrically. Values for defining quantities of garden ornaments or bridges or habitable architecture etc., provide a skeletal framework that can be transformed in a responsive manner with individual manipulations of values of $a \in A_i$. It essentially creates a framework in which the domain of a design space guides the number and relationship between the objects that constitute the resulting design parameters. It also lends itself well to the CAD-based parametric modeling paradigm Grasshopper 3-D, for which $\lfloor a \rfloor \in A_i = 10$ (where $a \in \mathbb{R}$) produces the following range for $f(a)$:

$$Ran\,(f) = \{\sigma_n \in \mathbb{R} \mid \forall \sigma_n : 0 < \sigma_n \leq 40\}.$$

This constraint has facilitated a greater ease of integration regarding the scaling of float values within Grasshopper 3-D. If the values of A_i are continually iterated by the designer, then aspects of the design process may be programmed to become somewhat automated such that the composition of a garden space is one concerned with the generation of a vast family of diverse typologies. By this I mean that the values of $a \in A_i$ can be controlled and adjusted by the designer individually among any number of distinct design phases, for which the totality of the design output (values of σ_n) is a direct response to this input, though because of the nature of $f(a) = \sigma_n$ and its various summations, such an output may produce novel or unexpected spatial configurations.

Figure 52: Perspective view of $c_6 \in \mathcal{G}_1$

A design then, and indeed the exploration of a *design space*, can become a performative action[68] where the *possibility* for locating myriad designs enhances design knowledge by the fact that some resultant forms may be initially unknown or at least unforeseen at the point of assigning values of $a \in A_i$. From this appropriation of $\mathfrak{B}(\mathbb{K})$, a definitive set of mappings of values is used to guide the creation of the theoretical landscape «*No-Stop Garden*». My initial design investigations have limited the absolute set of working landscape elements within «*No-Stop Garden*» to the objects {earth, moving water, still water, topography, habitable, architecture, bridges, paths, flora, rocks, garden ornaments}, though I envisage future design integration of the remaining objects of {exterior sounds, fauna}. Utilizing a small-scale GUI interface written in Pd, a series of seven sliders, each corresponding to $A_i \in \mathfrak{A}$ output float values (where $0 < a \leq 10$) whose particular sums are accorded with $f(a) = \sigma_n$ as described previously. This output is then piped via OSC (open sound control) directly to a series of computational components in the Rhinoceros 3-D plug-in Grasshopper 3-D. Grasshopper exists as a visual programming language within Rhinoceros 3-D in which compo-

68 R. Oxman, "Performative design: a performance-based model of digital architectural design," *Environment and Planning B: Planning and Design* 36/6 (2009): 1026–1037.

nents (lines, points, surfaces etc.) and routines (vectors, projections, rotations etc.) are defined in relation to each other and thus exist as algorithms. The controlling of individual parameters—through feeding data to individual components—can be achieved dynamically through GUI sliders, buttons etc., or other external interfaces via OSC or TCP (transmission control protocol). This paradigm essentially allows for a continuing morphing of the attributes of the architectonic model (that appears on output as Rhinoceros geometry) over time, or through multiple iterations or evolutions.

Conway's Rule 2,3/3 as Evolutionary Guide

The conversion of $\mathfrak{B}(\mathbb{K})$ into the piecewise $f(a) = \sigma_n$ has enabled a clearer nexus to become established between using the spatial relationships of Kyu Furukawa Teien as a holding framework for flushing through data. That data can be introduced and manipulated via GUI sliders enables a real-time tracking of the influence and parity of the individual parameters of $A_i \in \mathfrak{A}$ and how even subtle manipulations may have drastic effects on the output objects {earth, moving water, still water, topography, habitable, architecture, bridges, paths, flora, rocks, garden ornaments}. But to create more of a sense of an agent-based approach, as was the case in *Shin-Ryōan-ji*, the integration of a CA (cellular automata) is used as a method by which landscapes can not only "grow" from the input of GUI slider variables of $a \in A_i$ over successive generations, but also as a way to match the vastness of the set of architectural objects presented in «*No-Stop City*». John Conway's popular "game of life"[69] rule is used to drive the CA in «*No-Stop Garden*» primarily because of the potently chaotic nature of the rule. Archizoom's «*No-Stop City*» uses a set of four mirrors to generate an apparently infinite grid of the reflected single cell of the architectural model. I use this idea directly also in «*No-Stop Garden*», though implement Conway's "game of life" as a means to both stipulate and confine the dimensions of a grid, $\mathcal{P}_G$, and then to populate this grid with gardens on the cells (c_n) that are 'alive' (that is, $c_n = 1$). Rather than utilizing Archizoom's infinite grid, «*No-Stop Garden*» uses a 5×6 grid which is initially populated via a random seeding, then according to Conway's rule, successive gardens generated across a number of generations ($\mathcal{G}_t$) within the grid. This is essentially a function of my desire to present the possibility for *unknown* or novel forms of garden to emerge from the design process.

Conway's game of life rule 2,3/3 is rather simple, yet an excellent way in which to create a vast number of results quickly. The rule states that for an environment of cells the measure of its fitness (i.e., the number of cells that are alive or dead) at any point in its evolution over time is constrained by the number of neighborhood cells that are alive or dead.[70] Rule 2,3/3, here notated as $\boldsymbol{\rho}$, can then be expressed as:

69 M. Gardner, "Mathematical Games: The fantastic combinations of John Conway's new solitaire game 'Life'," *Scientific American* 233 (1970): 120-131.
70 C. Bays, "Introduction to Cellular Automata and Conway's Game of Life," in *Game of Life Cellular Automata*, ed. A. Adamatzky (London: Springer-Verlag, 2010), 1–7.

$$Let\ \boldsymbol{\rho} := \mathsf{E}_1, \mathsf{E}_2, \ldots / \mathsf{F}_1, \mathsf{F}_2, \ldots$$

Here, E_i is the environment, or number of live cells required to keep a living cell alive, and F_i is the fertility, or number of live cells required to bring a dead cell to life. Thus $\boldsymbol{\rho}$ dictates that 2 or 3 neighborhood cells are necessary to keep a functioning cell alive in the next generation, and if a cell is dead, for it to come to life, it will require 3 living cells around it to re-generate. Neighborhood cells are defined as those that share a boundary with the cell in question.

Figure 53: cell $c_7 \in \mathcal{G}_0$ containing garden L_7

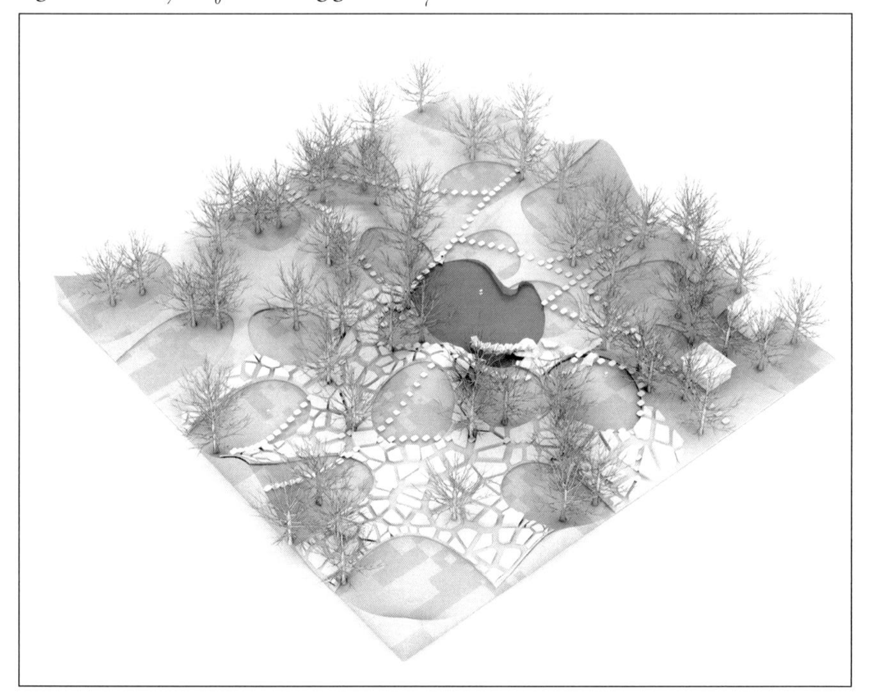

For «*No-Stop Garden*» the grid of cells under question is limited to a grid of 5 × 6, though traditionally, CA simulations are often constituted over a far greater area. Starting at generation $t = 0$ with the power set of the 5 × 6 grid, $\mathcal{P}_G$, a pseudo-random number generator (denoted here as $\boldsymbol{\kappa}$) is applied to all cells to derive values of $c_n \in R_0$ such that $\forall c_n \in R_t \rightarrow c_n \in \{0, 1\}$. To obtain the set of all 'active cells,' $\mathcal{G}_t$ ($\forall c_n = 1$) for generation $t \geq 0$ the compliment to $\mathcal{P}_G$ is found such that $\mathcal{G}_t = R_t \setminus \mathcal{P}_G$. The following diagrams give an account of the procedure:

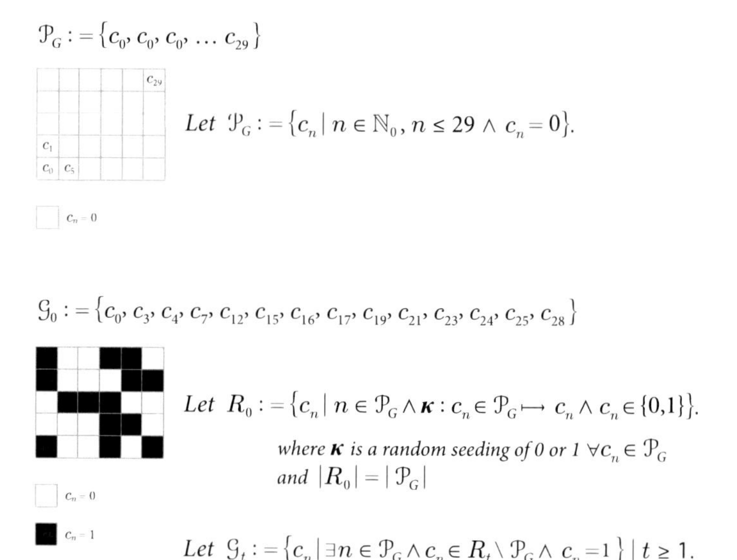

In the above example, fourteen black squares show the initial generation $\mathcal{G}_0$ of cells c_n that will be hosts to garden spaces. «*No-Stop Garden*» is initiated then via a random seeding of $\mathcal{P}_G$ to firstly find which cells are 'alive' and active (i.e. contain a value of 1) and therefore which can be used to *plant* a garden. The cells for which $c_n = 1$ thus also represent containers where the each garden's attributes are complied such that:

Garden generation :

$$\left(L_\alpha := \bigcup_{i=1}^{7} A_i^\alpha \middle| \; \alpha = (n \in \mathcal{G}_t)\right) :\Longleftrightarrow \left(\exists c_n \in \mathcal{G}_t = 1\right).$$

$$n \in \mathcal{G}_0 = \{0, 3, 4, 7, 12, , 15, 16, 17, 19, 21, 23, 24, 25, 28\}.$$

Here, L_α represents the union of values of $a \in A_i^\alpha$ only as a consequence of the existence of active cells ($c_n = 1$) in $\mathcal{G}_t$. The index α is used to differentiate between values of a among the fourteen different gardens of $\mathcal{G}_0$. This framework is then followed through via the function $f(a) = \sigma_n$ such that the set of counts of garden objects in $\mathfrak{O}$ can be further computed. In the process then of designing «*No-Stop Garden*», random (or alternatively aesthetically *controlled*) values of $a \in A_i^\alpha$ are seeded after firstly deriving $\mathcal{G}_0$, for which subsequent manipulations of values of a (via the Pd GUI slider interface) enable a level of intervention from the designer regarding the overall aesthetics of each of the resultant 14 initial garden spaces (L_α). The completion of this initial process then allows the cells of $\mathcal{G}_0$ to potentially store information (values of a and σ_n) regarding various spatial attributes of the gardens. The process of 3-D form generation is actuated within the Rhinoceros 3D plug-in Grasshopper 3D.[71] The plug-in essentially interprets float values of σ_n by using components that assign both spatial locations within a cell for trees, architecture, rocks and topographic features etc., as well as trajectories for paths and orientation of water features (see Figure 53 for an example garden). The automation and means by which the program assigns such locations is fixed, though to facilitate a greater diversity across the final designs some random data generation is used for 2-D/3-D point-cloud generation, list value culling, random seeding, and to drive non-repetitious structures like Voronoi tessellation procedures,[72] the Boids flocking algorithm[73] and Substrate patterning algorithm.[74] The overall approach then is to utilize values of σ_n in a structured yet meaningful way by applying simple transformations of the values through re-mappings or scaling within Grasshopper 3D. As such, «*No-Stop Garden*» is a highly stylized construct in which features, such as the shape and size of garden elements (e.g., paths, trees, architecture, garden ornaments etc.) reoccur throughout all the gardens as spatial constants for which different configurations, according to values of σ_n, highlight emerging typologies within L_α.

The Inheritance from Neighborhood Gardens

When running a CA simulations over t generations, the initial randomly seeded binary values for the starting grid, combined with the particular rule applied are the only factors in determining the future values of each cell. Given the greater complexity of «*No-Stop Garden*» as a collection of multiple landscape elements, and in particular the many possible values of a for governing the resultant spatial dimensions of objects within the garden, I used a more literal interpretation of the fertility aspect of Conway's rule as a way to produce a more meaningful inheritance framework. Looking more closely at the first generation of gardens $\mathcal{G}_{t=1}$ the application of $\boldsymbol{\rho}$ (Conway's rule 2,3/3) unfolds similarly as with $\mathcal{G}_{t=0}$. The set $R_{t\geq1}$ is derived from R_0 via an application of $\boldsymbol{\rho}$, and is thus distinct through its use of $\boldsymbol{\rho}$ rather than $\boldsymbol{\kappa}$. The set $\mathcal{G}_1$ then contains values of $c_n = 1$ and is derived through $R_1 \setminus \mathcal{P}_G$:

71 Arturo Tedeschi, *Parametric Architecture with Grasshopper* (Brienza: Le Penseur, 2011).

72 F. Aurenhammer, "Voronoi Diagrams–A Survey of a Fundamental Geometric Data Structure," *ACM Computing Surveys* 23/3 (1991): 345-404.

73 C. Reynolds, "Flocks, herds and schools: A distributed behavioral model," *Computer Graphics* 21/4 (1987): 289–296.

74 M. Ogawa and K. L. Ma, "code_swarm: A Design Study in Organic Software Visualization," *IEEE Transactions on visualization and computer graphics* 15/6 (2009): 1097–1104.

$$\mathcal{G}_1 := \{c_3, c_6, c_7, c_{12}, c_{15}, c_{19}, c_{24}, c_{27}, c_{28}, c_{29}\}$$

$$\mathcal{N}_1 \subset \mathcal{G}_1 = \{c_6, c_{27}, c_{29}\}$$

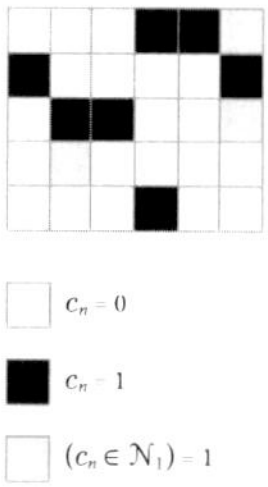

$c_n = 0$

$c_n = 1$

$(c_n \in \mathcal{N}_1) = 1$

Recall $R_0 := \{c_n \mid n \in \mathcal{P}_G \wedge \boldsymbol{\kappa} : c_n \in \mathcal{P}_G \mapsto c_n \wedge c_n \in \{0,1\}\}$.

Let $R_{t\geq 1} := \{c_n \mid \exists n \in \mathcal{P}_G \wedge \boldsymbol{\rho} : c_n \in R_0 \mapsto c_n \wedge c_n \in \{0,1\}\}$.

where $\boldsymbol{\rho}$ *is Rule 2,3/3 applied to* $c_n \in R_0$ *and* $|R_0| = |R_{t\geq 1}|$

Let $\mathcal{N}_t := \{c_n \mid \exists n \in \mathcal{G}_t \wedge c_n \in \mathcal{G}_t \setminus \mathcal{G}_{t\text{-}1} \wedge c_n = 1\} \mid t \geq 1$.

The set $\mathcal{N}_1$ is a subset of $\mathcal{G}_1$ and is the set of 'new' cells to appear in generation $t = 1$: i.e., those cells initially dead in $\mathcal{G}_0$ though because they had three living neighborhood cells are brought to life in $\mathcal{G}_1$. The light grey cells represent unchanged cells (i.e., cells that are still active) from $\mathcal{G}_0$. Rather than simply assigning another set of random or considered values of a to generate L_6, L_{26}, L_{29}, I developed an approach that utilizes the notion of the inheritance of attributes from the surrounding gardens in the previous generation, that is, $L_\alpha \in \mathcal{G}_0$. Firstly, a new set $R_{t\geq 1}$ describes the set of c_n that are a result of the application of $\boldsymbol{\rho}$ on R_0. The set of active cells for $\mathcal{G}_{t=1}$ is thus derived from the relative compliment of $R_1 \in \mathcal{P}_G$ and the set of new cells collected within $\mathcal{N}_1$ is similarly the relative compliment $\mathcal{G}_t \setminus \mathcal{G}_{t\text{-}1}$. A *Garden inheritance* condition is used for finding and guiding the computation of values of a via the *Garden evolution* rule. These two frameworks are based on the formal definition of the two sets, $\mathcal{J}_t$ and $\mathcal{M}_t^\alpha$ for which $\mathcal{J}_t \supset \mathcal{M}_t^\alpha$:

Let $\mathcal{J}_t := \{c_n \mid \exists n \in \mathcal{G}_{t\text{-}1} \wedge c_n I_\mathsf{F} \, (\exists! c_n \in \mathcal{R}_t \cap \mathcal{P}_G) \wedge c_n = 1\} \mid t \geq 1$.

where I_F *is an incidence relation such that* c_n *satisfies* F_1 *of* $\boldsymbol{\rho}$

Let $\mathcal{M}_t^\alpha := \{c_n \mid \exists n \in \mathcal{G}_{t\text{-}1} \wedge c_n I_\mathsf{m} \, (\exists! c_n \in \mathcal{N}_t) \wedge |\mathcal{M}_t^\alpha| = 3 \wedge c_n = 1\} \mid t \geq 1$,

$\alpha = n \in \mathcal{N}_t$, $\mathcal{M}_t^\alpha \subset \mathcal{J}_t$

where I_m *is an incidence relation such that* $c_n \in \mathcal{M}_t^\alpha$ *are neighbouring cells to* $\exists! c_n \in \mathcal{N}_t$

Here, the set $\mathcal{J}_t$ (for $t \geq 1$) is the set of c_n that are required to allow $(c_n \in \mathcal{G}_0) = 0$ to map to $(c_n \in \mathcal{G}_1) = 1$. This is formally dictated via $\boldsymbol{\rho}$, and in particular F_1 in $\boldsymbol{\rho}$. The notation I_F is used to describe this condition.

Figure 54: cell $c_3 \in \mathcal{G}_0$ and $c_4 \in \mathcal{G}_0$ containing garden L_3 & L_3

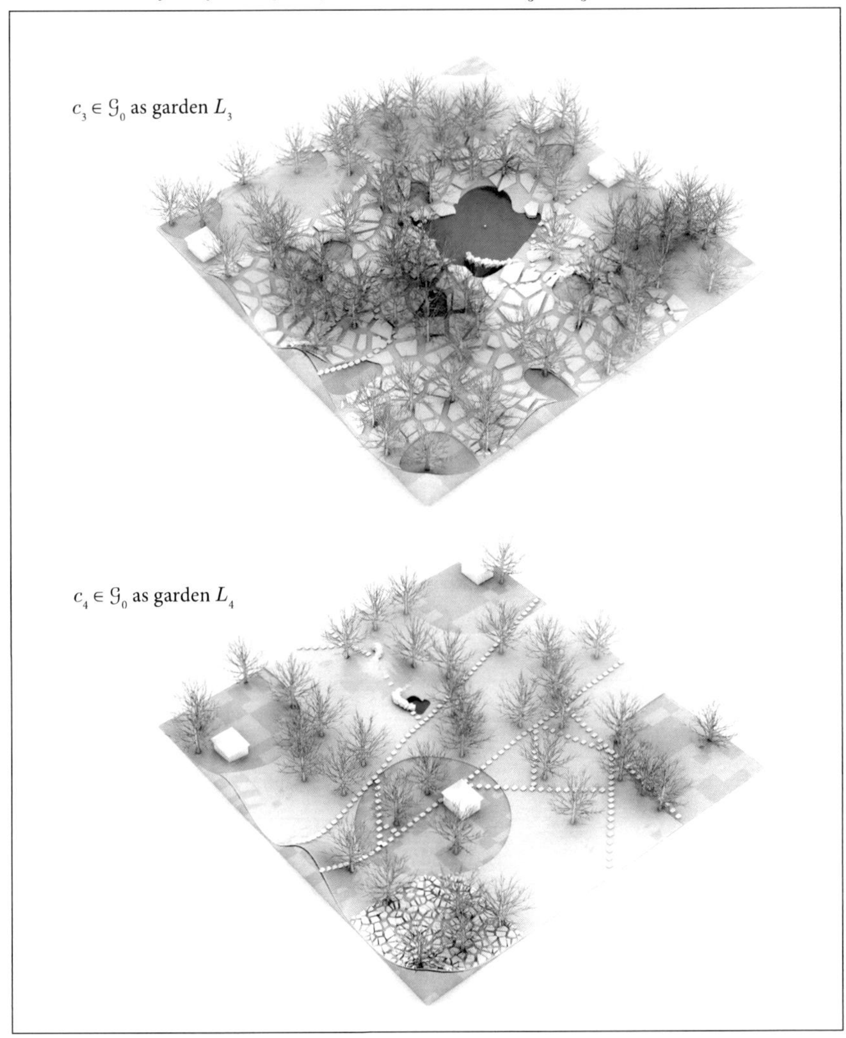

From these sets, the rule of inheritance and the method for finding values of a in the gardens tagged to $c_n \in \mathcal{N}_t$ follows as:

Garden inheritance:

$$\left(\exists c_n \in \mathcal{J}_t = 1 \Leftrightarrow c_n I_{\mathsf{F}}(c_n \in \mathcal{R}_t \cap \mathcal{P}_G)\right) :\Longleftrightarrow \left(\mathcal{M}_t^{\alpha} \subset \mathcal{J}_t \Leftrightarrow c_n \in \mathcal{M}_t^{\alpha} I_{\mathsf{m}}(\exists! c_n \in \mathcal{N}_t)\right)$$

Garden evolution:

$$\left(L_\alpha := \frac{1}{|\mathcal{M}_t^\alpha|}\sum_{n\in\mathcal{M}_t^\alpha}^{3} \mathrm{D}_n \;\middle|\; \alpha \in \mathcal{N}_t\right) :\Longleftrightarrow \left(\exists c_n \in \mathcal{G}_t \setminus \mathcal{G}_{t-1} = 1 \wedge t \geq 1\right).$$

D_n *row matrix:*

$$\left(\mathrm{D}_n := \left[\, a_1^\alpha, a_2^\alpha, a_3^\alpha, a_4^\alpha, a_5^\alpha, a_6^\alpha, a_7^\alpha \right] \;\middle|\; n \in \mathcal{M}_t^\alpha\right) :\Longleftrightarrow \left(a_i^\alpha \in A_i^\alpha \Leftrightarrow \bigcup_{i=1}^{7} A_i^\alpha\right)$$

The *Garden inheritance* is a condition that states that if there exists a $c_n = 1$ in the set $\mathcal{J}_t$ then it is a result of $F_1 \in \boldsymbol{\rho}$, and if this is true, then $\mathcal{M}_t^\alpha$ is a subset of $\mathcal{J}_t$ and are neighborhood cells to exactly one c_n in $\mathcal{N}_t$. The *Garden evolution* rule thus operates similarly to the previous *Garden generation* rule, though here is implemented according only to the existence of the set $\mathcal{N}_t$ $(\mathcal{G}_t \setminus \mathcal{G}_{t-1})$ for $c_n = 1$. To obtain values of $a \in L_\alpha$ the previous union of A_i^α is transformed via the D_n *row matrix*, such that $\forall a \in A_i^\alpha$, D_n is considered a single row matrix for the facilitation of the averaging of $a \in L_\alpha$ via c_n that satisfy I_F. This novel re-application of $\boldsymbol{\rho}$ is enacted as a means to obtain values for $a \in L_\alpha$ that are a function of the neighboring context of gardens. This then allows for a sense that the successive generations of new gardens to appear will evolve to some degree through a dependency of previous values of a within already established gardens within the grid. For $L_6 \in \mathcal{G}_1$, this process is initiated via the cells c_6, c_{27}, c_{29} and essentially equates to the following:

$$\begin{aligned} \textit{if } \mathcal{G}_1 &= \{c_3, c_6, c_7, c_{12}, c_{15}, c_{19}, c_{24}, c_{27}, c_{28}, c_{29}\}, \\ \mathcal{J}_1 &= \{c_0, c_7, c_{12}, c_{21}, c_{23}, c_{24}, c_{28}\}, \\ \mathcal{M}_t^\alpha &= \{c_0, c_7, c_{12}\}, \\ \mathcal{N}_1 &= \{c_6, c_{27}, c_{29}\}, \end{aligned}$$

$$\textit{then } \{c_0, c_7, c_{12}\} \in \mathcal{G}_0 \qquad\qquad c_6 \in \mathcal{G}_1$$

$$\longrightarrow$$

$$\begin{aligned} L_6 &= \frac{1}{|\mathcal{M}_t^\alpha|}\sum_{n\in\mathcal{M}_t^\alpha}^{3} \mathrm{D}_n \\ &= \frac{\mathrm{D}_0 + \mathrm{D}_7 + \mathrm{D}_{12}}{|\mathcal{M}_1^6|} \\ &= \left[\frac{(a_1^0 + a_1^7 + a_1^{12})}{|\mathcal{M}_1^6|}, \frac{(a_2^0 + a_2^7 + a_2^{12})}{|\mathcal{M}_1^6|}, \ldots \frac{(a_7^0 + a_7^7 + a_7^{12})}{|\mathcal{M}_1^6|}\right] = \left[a_1^6, a_2^6, a_3^6, \ldots a_7^6\right] \end{aligned}$$

This approach pivots on finding the active neighborhood cells to c_6 in $\mathcal{G}_0$ so as to average their a values for generating L_6. Thus $(\mathcal{M}_1^6 \subset \mathcal{J}_1) \in \mathcal{G}_0$ for which $\{c_0, c_7, c_{12}\} I_m c_6$, hence $c_6 \in (\mathcal{N}_1 \cap \mathcal{G}_1)$. Because «*No-Stop Garden*» operates within a 5 × 6 grid there is a finite, relatively small set of neighborhood cell combinations (the set $\mathcal{N}_t$, where $|\mathcal{N}_t| = 3$) that will allow a 'dead' cell to come alive in generation $t \geq 1$ due to $\mathsf{F}_1 \in \boldsymbol{\rho}$. Figure 55 summarizes these neighborhood cell conditions.

Figure 55: Generalization of neighborhood cells in $\mathcal{P}_G$

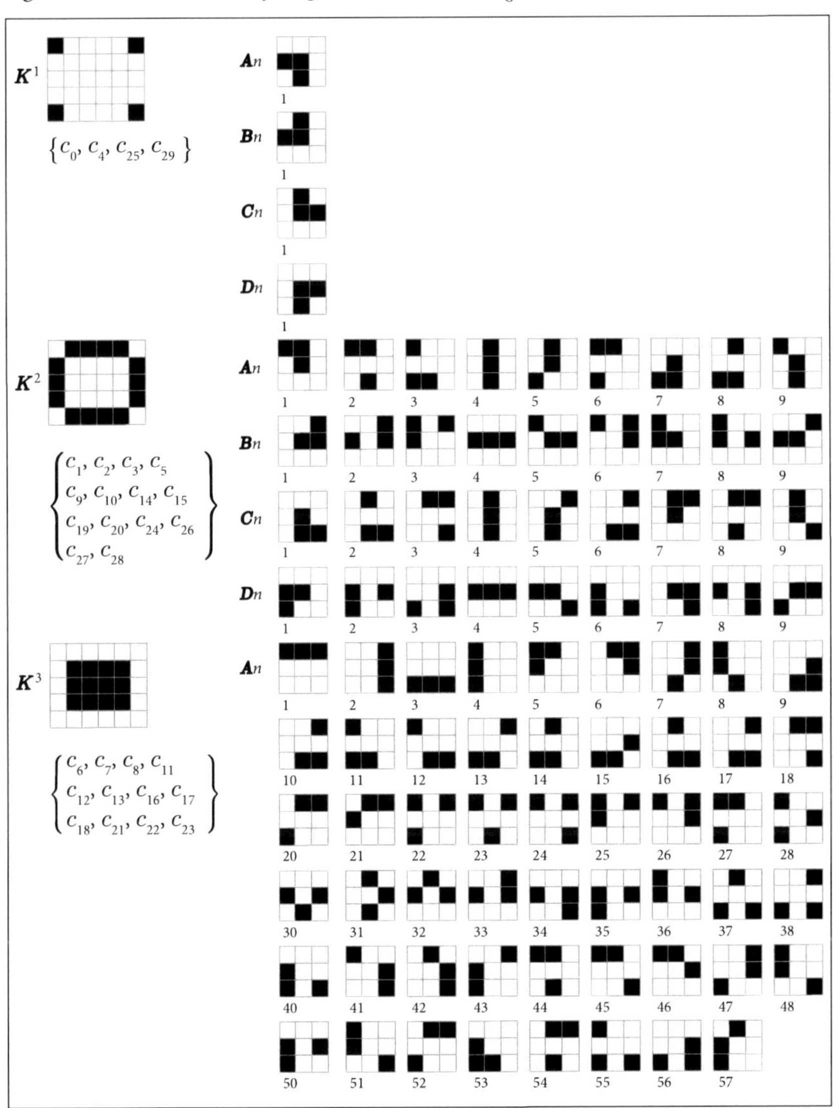

The sets K^n account for different configurations of $(c_n = 0) \in R_t$. These are the classes of possible 'dead' cells that require 3 neighboring active cells (combinations shown in A*n*, B*n*, C*n*, D*n*) to come to life in the next generation. For example, from Figure 55, the case of L_6 falls under K^3-A19, while the other members of $\mathcal{N}_1$ are from K^1 and K^2, namely:

$$\mathcal{N}_1 = \{c_6, c_{27}, c_{29}\}$$

K^2 - $C8$:

$$\{c_{21}, c_{23}, c_{28}\} \in \mathcal{G}_0 \qquad c_{27} \in \mathcal{G}_1$$

K^2 - $C1$:

$$\{c_{23}, c_{24}, c_{28}\} \in \mathcal{G}_0 \qquad c_{29} \in \mathcal{G}_1$$

The parallel coordinates plot in figure 56 shows the actual values for $a \in L_\alpha$ for the gardens housed on $c_n \in \mathcal{N}_1$ in «*No-Stop Garden*». The results of this approach can be observed in the actual designed gardens as shown in Figure 57. Though I have explored only the first generation of gardens here, the possibility for an infinite number is plausible. In any case, what I believe is pertinent in «*No-Stop Garden*» is that which Dennett[75] nominates as the concept of *vastness*, i.e., the inconceivably large dimensions of a design space, for which in my case, the exploration of its possibilities are greatly enhanced by a computer-generated semi-autonomous routines. What the designer sets in motion then is limited to the considered or random values of a created for all cells active in $\mathcal{G}_0$. The diversity of the ensuing generations of $\mathcal{G}_n$ may offer up any number of variations in landscape form that will continually become iterated and combined.

Figure 56: Parallel coordinates plot of $a \in L_\alpha$

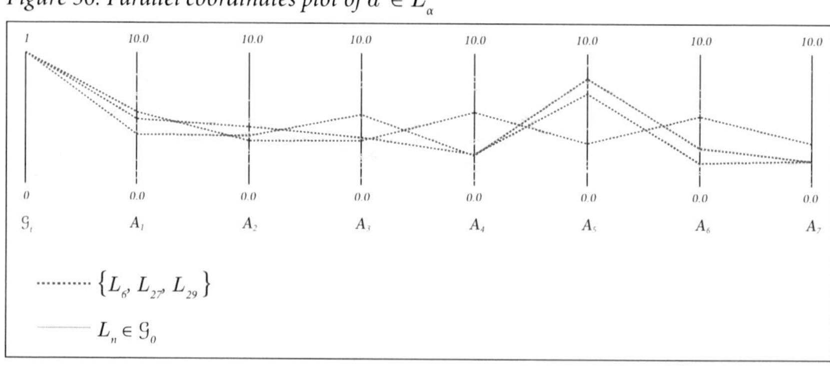

75 D. C. Dennett, *Darwin's Dangerous Idea: Evolution and the Meanings of Life* (New York: Simon & Schuster, 1995).

Figure 57: Complete iterations of gardens

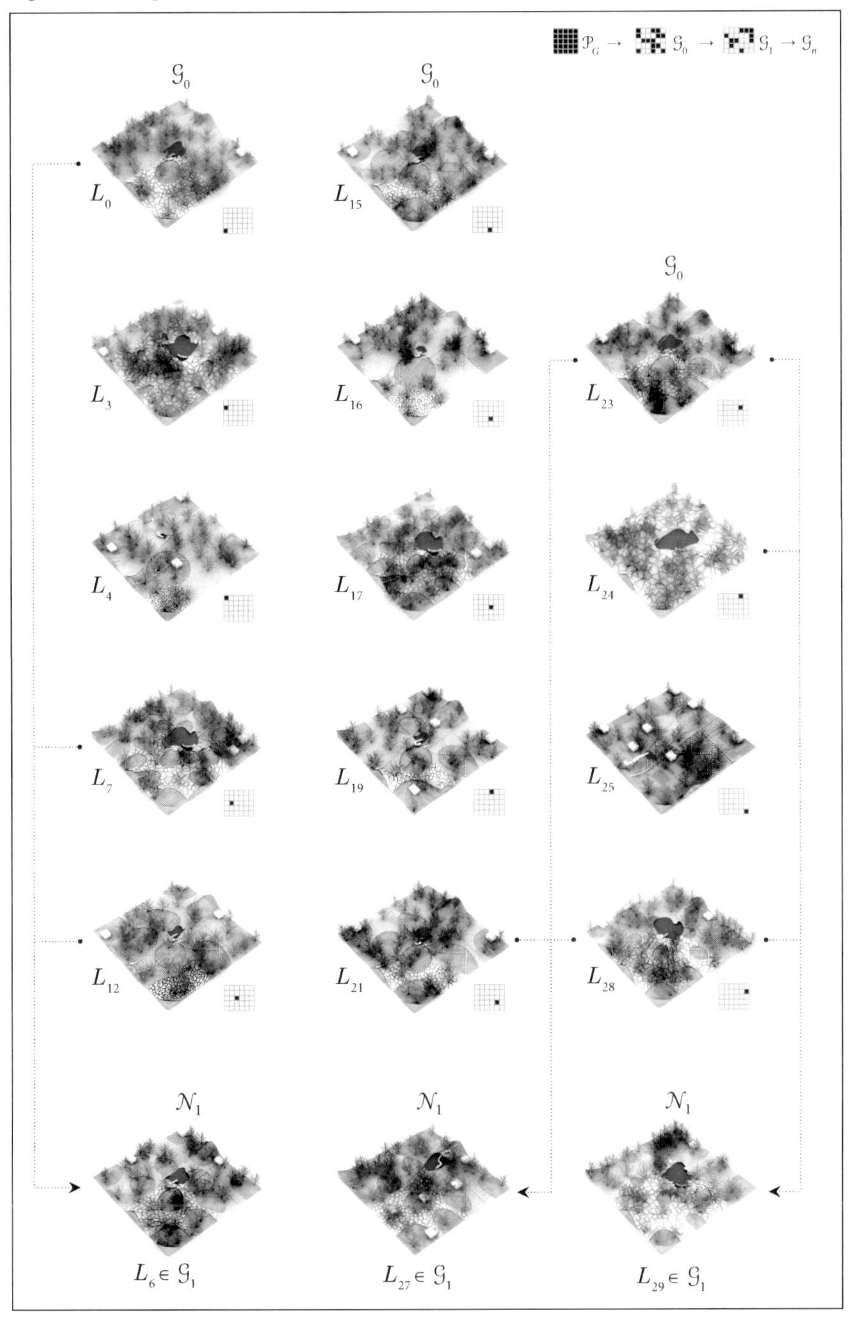

On Design Space Exploration

Perhaps the link between the architectonic form finding explored in *Kyu Furukawa Teien Shūsaku* and the design space exploration of «*No-Stop Garden*» is one forged through an emphasis on the process of spatial generation as an act that is bound within a framework that demands repetition or repetitious production. Indeed, this is a similar position as to how Aureli describes Archizoom's «*No-Stop City*» in which:

"the city [is considered] as a continuous system rather than a collection of objects. Foreshadowing later theories of media and immaterial production, Branzi [Archizoom's lead architect] emphasized that if the city was integrated into the cycle of production, then producing it was only a matter of programming, not designing, its built structures" [76]

But for «*No-Stop City*» the model of the city is not utopian in the sense of *New Babylon* (1953) of Constant Nieuwenhuys. Here, and as Mark Wigley notes, the "utopia [of *New Babylon*] is a picture of society that ignores material conditions, an idealization of reality."[77] Instead, for Archizoom, the city becomes a crucible for criticality, for critical thinking and for questioning the impacts and goals that a utopic vision of architecture may deliver. It is too a model of global influence and of urbanization such that

"No-Stop City is a critical utopia, a model of global urbanization in which design is conceived of as a tool for modifying the quality of life and territory. The city presents the same organization as a factory or a supermarket. Interior spaces in No-Stop City, with air-conditioning and artificial lighting, allow city dwellers to organize new typologies of open and continuous inhabitation, intended for new forms of association and community."[78]

This emphasis in «*No-Stop City*» on programming within a series of constraints is certainly a shared goal between all three projects that I have explored in this chapter. But «*No-Stop Garden*» and *Kyu Furukawa Teien Shūsaku* share an even more specific methodological approach. Here, a purely theoretical exploration of a design space is utilized as a means to both visualize an exploration of a small area of the *spatial continuum* as well as highlight the framework for which a specific end design can be a result of selective procedures such as culling or optimizing. Here then is the essence of what Frédéric Migayrou has described as *non-standard architecture*, where:

76 P. V. Aureli, *The Project of Autonomy: politics and architecture within and against capitalism* (New York: Princeton Architectural Press, 2008), 77.
77 Mark Wigley, *Constant's New Babylon: They Hyper-Architecture of Desire* (Rotterdam: Witte de With, 1998), 235.
78 Alison, Brayer and Migayrou, *Future City*, 276.

"digital tools and their capacity for algorithmic calculation allow one to enter on a solid footing into thc domain of a continuous formal schematism revolutionizing the logic of architectural design. Here, form becomes a morphogenetic a priori, the form chosen to embody architecture being in a state for defining a singularity only in a continuum in perceptual evolution."[79]

Indeed the logic of *«No-Stop Garden»* and *Kyu Furukawa Teien Shūsaku* has been exactly to approach form generation as not the fixing of a singularity and its materiality or formal geometry through an unchanging series of relationships, but rather, as a means to move from stasis to a state that Greg Lynn would define as curvaceous[80]—that is, as an expression of calculus where a state of change defines the *progression* rather than the fixed point. But this notion of the continuum that encompasses those formal possibilities that each of the projects may take, is also how Woodburry and Burrow conceive of design space. In particular, they note the ramifications of designs locatable within real-world constraints:

"designs themselves are important objects, irrespective of what they represent. That they are placed in the design space captures a position from which further exploration can be made. This is a corollary of the fact that accessibility is the measure of possibility: designs without physical interpretation or with poor qualities may be the basis for other realizable designs."[81]

But what has guided both *«No-Stop Garden»* and *Kyu Furukawa Teien Shūsaku* in the exploration of their respective design spaces has not so much been a searching out for *optimizations* of particular qualities of a design, but rather, in the aesthetics of Cage, an exploration of indeterminacy and the power of uncertainties to generate new knowledge. This is akin to the notion Cage had of the eschewing of a set identity for a work of art. As Thomas DeLio has noted regarding works such as *Variations II*: "Cage has made available all the possible variants of one type of structure but has, himself, singled out none in particular to be the specific form of the work."[82] This is a similar case as to both *«No-Stop Garden»* and *Kyu Furukawa Teien Shūsaku* where though multiple end forms are presented, their identities—in those traditional musical terms—emerge through the very fixed and unchanging processes that give rise to their spatial continuum or design space. In essence, both projects presents themselves as "open forms" from the very beginnings of the design space exploration in that they are structured in a manner that conceals their resultant end forms to be represented only as statistical potentials rather than fixed tangible objects. At the beginning of the

79 Frédéric Migayrou, "non-standard orders: 'nsa codes', " in *Future City*, eds. Jane Alison, Marie-Ange Brayer, Frédéric Migayrou and Neil Spiller (London: Thames and Hudson, 2006), 29.
80 Grey Lynn, *Animate Form* (Princeton: Princeton Architectural Press, 1999).
81 Burrow and Woodbury, "Whither Design Space?," 67.
82 DeLio, "John Cage's 'Variations II', " 369.

form finding process what is known of the end forms are only the dimensions of the continuum that they will be born from. In the case of *«No-Stop Garden»* parametric modeling has allowed this exploration of the uncertainties to become automated such that generating form is a means of speculating on the flow and combination between parameters and how these might cajole or resist integration into the whole. By this I mean that the structure of the process allows for a seemingly limitless approach to form finding, and one in which a vast variety of forms may be equally legitimate. The system then can be regarded as an experimental one in that the outcome is not a concrete one that is known, but can only be theorized within a statistical model. This is perhaps what Luciana Parisi notes as the power of parametric approaches to create a state of fuzzy logic in the design process in which yes or no states (or even value judgments) are abandoned in favor of the mapping of uncertainties:

"Parametric control is marked by the ingression of immanent speculation into rationality: an infection with abstraction that irreversibly drives all forms of decision making beyond yes and no states. In short, parametric control is a mode of power that operates through the digital scripting of uncertainties."[83]

Thus, rather than relying on the infinite repetition of a single architectonic or other spatial elements to define the limits of a design space, *«No-Stop Garden»* iterates a series of static mathematical functions on a vast scale, that while creating a seemingly repetitious field of designs, also allows for subtleties of change and evolution to emerge within each repetition. This is also true of *Kyu Furukawa Teien Shūsaku* such that for both projects, the processes are completely self-contained and have an inner logic as to their methods (that is, they are not completely randomized in every respect) yet the resultant end forms similarly point beyond themselves and to the potential of further iterations. They are then not dissimilar to how Grey Lynn conceives of his notion of *inflection* as a method for integrating a number of forces for producing geometric contingencies. Lynn's spatial continuum is based on the mathematical constructs of differential equation modeling though it conceptual base is familiar one:

"Producing a geometric form based on a differential equation is problematic without a differential approach to series and repetition. There are two kinds of series, one discrete, the one of repetitive series, and the other continuous, the one of iterative series. The difference between each object in a sequence is an individuated state critical for each repetition."[84]

This individuated state within the models of *Kyu Furukawa Shūsaku* and *«No-Stop Garden»* allow the repetitions to continuously fill the spatial continuum, while also

83 Luciana Parisi, *Contagious Architecture: Computation, Aesthetics, and Space* (Cambridge (MA): MIT Press, 2013), 153.
84 Lynn, *Animate Form*, 33.

allowing for the variety between iterations to be traceable or identifiable. But because of this, the focus purely on form generation processes also pushes them beyond the traditional architectural model and into Migayrou's notion of the non-standard. Unlike true architectural models or architectural simulations they are non-specific in regards to site or program. They are purposely generic and exist as architectonic forms designed not as inhabitable spaces but contemplative objects occupying a threshold. This is also particularly true for *NURBS Map I / II* in which the traditional relationship between the *prescriptive* and the *descriptive* in architecture—as exemplified through the relationship between the plan and the diagram—is here interrogated and problemitized. In architecture, the plan is a written structure to which those constructing the object carry out actions. Materials, proportion, scale and dimension are predetermined and parsed into a notated system that has traditionally abstracted the intended three-dimensional object into a flat rendering of its salient spatial qualities. The diagram, as a foil to the plan, is similarly abstracted, though more often used as a projection of the object's narrative, layers of inquiry, or underlying forces usurped for the design process, and acts as a reflective or contemplative device that disregards its own materiality yet communicates ideas and solutions.[85] Often, architectural diagrams are not readily understood by an outside reader without a knowledge of the visual nomenclature of architectural diagrammatics—which themselves are usually project-specific. If the plan communicates the physical dimensions and intended materiality of the object to be made in tangible terms, the diagram describes the intangible: anything from a project's spatial aesthetics (mapped in visual terms) to its potential typology, morphology or even the ontology of the surrounding ideas that have driven, or will drive, the design process.

The three architectonic projects outlined in this chapter thus sit at the threshold between the plan and the diagram, and existing as models of spatial flux, generative action or mappings of spatialized musical data makes them distinctly proto-architectural. But their nature as interstitial objects between functional architectural form and discrete analytic visualizations also creates a potential for them to move beyond the nexus they occupy. This is perhaps what Heinrich Wölfflin describes when he speaks of the meaning of the open form within art:

"What is meant is a style of composition which, with more or less tectonic means, makes of the picture a self-contained entity, pointing everywhere back to itself, while, conversely, the style of open form everywhere points out beyond itself and purposely looks limitless."[86]

Of course the larger process by which these projects have formed also concerns my desire to represent the underlying structural proclivities of the Japanese garden at

85 Yi-Luen Lo and Mark D. Gross, "Thinking with Diagrams in Architectural Design," *Artificial Intelligence Review* 5 (2001): 135–149.
86 Heinrich Wölfflin, *Principles of Art History* (New York: Dover, 1932): 161 .

Kyu Furukawa Teien. In the case of «*No-Stop Garden*», FCA provided an analytical construct for creatively going beyond its original intention in an effort to usurp its structure for design automation, thus allowing for a method in which the garden is conceptualized as a series of mathematical relationships which are then re-applied in a new design space. The fundamental structures inherent in the garden as thus transmediated into controlling new forms. But this usurpation of underlying spatial principles or in the case of *Kyu Furukawa Teien Shūsaku*, of underlying musically meaningful information, has been for the usability of this information into a digital design environment where transmediation strategies can play out in more non-linear ways. As examples of what I consider *spatial thinking* they represent the paradigm of transmediation that has been explored in this chapter within both the auditory and architectonic realm where spatial information from the Japanese garden is collected, analyzed and abstracted so that it can be re-used within a framework of creative iteration.

Reflective Practice

On Spatial Thinking

The work presented in this book has not only argued that the Japanese garden represents a type of spatial exemplar in terms of soundscape and landscape qualities, but that the creative and analytical work undertaken has served to form a foil or companion to the precedent created by composer and philosopher John Cage. But indeed the notion of Marjorie Siegel and the framework for enacting myriad types of transmediations between media is not an approach only used by Cage or one that can be exclusively traced to the Muromachi landscape painter and garden maker Sesshū Tōyō, though for this study they represent the most direct and influential precedents that have shaped the investigation. Indeed the concept of *re-representing* information from one realm into another can also be traced to the sixth-century Roman philosopher Boethius, whose classification of music in his treatise *De Musica* provides an insight into those more universal themes occurring in numerous later manifestations of art that represent the structures of the natural world through sound.

For Boethius, artistic expression regarding the discipline of music (*musica universalis*) could be categorized as falling within five sub-groups: *Musica mundana* (music of the spheres), *Musica humana* (music of the human body), *Musica instrumentalis* (instrumental music as well as the human voice), and *Musica divina* (music of the gods). The concept of the "music of the spheres" is attributed to the earlier Greek philosopher and mathematician Pythagoras, whom through the influential writings of Plato later became disseminated to medieval thinkers. Pythagoras and his followers understood that musical intervals could be expressed as ratios (e.g. subdividing a monochord as 1:2 gives an octave, 2:3 a perfect 5th, 3:4 a perfect fourth etc.), and as such, that mathematics becomes a mediator between phenomena in the world and the human condition:

"The Pythagoreans taught the analogy of mathematical (and thereby harmonic) proportions: e.g. 2:4 = 3:6. Plato employed the idea of analogy in a more general sense when, in his *Timaeus* and *The Republic*, he related the world of ideas analogously to the world of matter. Medieval Neo-Platonism held that music is an analogy to all ordered being [. . .] From this understanding derived the discipline of *musica speculativa*, in which music was understood to serve as a mirror for the remainder of the world. Consequently, the first postulate of speculative music is that tones, like numbers, are ontologically prior to material existence."[1]

The notion of the ancient Greeks regarding mathematics as a mediator between the abstract and the actualized and that music was an enabling device that revealed "the remainder of the world" or that which is currently hidden from view, captures the modus operandi of both Cage's *Ryoanji* precedent and my own creative works that have sought to extend Cage's method. But rather than being limited to the notion of music, I frame my own approach as based in sound, though driven by an exploratory or speculative interdisciplinary context.

For later Baroque scholars though, this powerful concept of abstract representation further enabled a means in which to understand the movement of heavenly bodies as an unheard musical event. In Johannes Kepler's 1619 study of the proportions of the natural world, *Harmonices Mundi* (The Harmony of the World),[2] the resultant third law of planetary motion is a function of his desire to prove that the ratios and proportions that govern the movement of the planets around the sun are analogous to the ratios used to generate *musica instrumentalis*:

"As Keller emphasized many times, the laws [of planetary motion] are a mere tool and as such subservient to his nobler goal: to demonstrate the identity of order in nature, in the psycho-physical disposition of humans, and in music; to prove that the age-old belief in a universe resonating in accordance with musical consonances can be corroborated with the help of 17th-century physics; and to show that there are reasons to trust that the same harmonious proportions pertain to innumerable other aspects of the universe."[3]

It is perhaps this model of the connection between the pursuits of knowledge within the scientific *and* the artistic realm that provides a more complete frame for the work I have presented in this book regarding the various interrogations into the Japanese garden as an exemplar spatial entity. Under what I have termed then *spatial thinking*, the historical precedent of such a practice, and in particular, the complimentary approaches of Pythagoras, Boethius and Kepler can thus be regarded in the same fashion as that which William Owen describes as the function of *information maps*[4] Here, such maps are models or graphic representations of an intangible source. In-

1 Siglind Bruhn, *The Musical Order of the World: Kepler, Hesse, Hindemith* (Hillsdale, NY: Pendragon Press, 2005), 18.
2 Johannes Kepler, *The Harmony of the World*, trans. E. J. Aiton, A. M. Duncan and J. V. Field (American Philosophical Society: Philadelphia, 1997).
3 Bruhn, *The Musical Order of the World*, 13.
4 William Owen, ed. "I saw a man he wasn't there," in *Mapping* (Mies: RotoVision, 2005), 154.

deed, in terms of the Japanese garden as a source for sampling, or as an ostensive object or even autonomous agent, it can be of course readily understood in those basic terms that identify its tangible elements—plants, trees, water, rocks, sounds etc. But that which forms its underlying structure and multi-sensory inter-relationships, its deeper spatial proclivities and spatial tensions, these elements are akin to the "hidden" proportions that describe the distance ratios between the planets or the harmonic proportions that create sounding vibrations. As I have examined throughout this book, the geometries of Ryōan-ji or Sesshutei, the relationship at Kyu Furukawa Teien between *soto* and *uchi*, dry and wet, *yin* and *yang*, or the weaving of soundscape and landscape elements at Koishikawa Korakuen, or the use of *shakkei* at Adachi represent some of those deep-level relationships that combine to produce particular rarefied multi-sensory experiences within these gardens. *Spatial thinking* then is a process by which these underlying identities, these traits and frames of knowledge are unearthed and examined not only through the lens of observation as a act of revealing for erudition, but similarly as an act to enable further artistic production. I have outlined in this book a number of objective and at times highly technical methods for the analysis of the spatial properties of the Japanese garden that perhaps approach a similar ground to the work of those early philosophers, Medieval and later Baroque scientists that I have already mentioned. But whether my own approach has been facilitated through the use of mathematical constructs for describing relationships in Kyu Furukawa Teien using FCA, digital modeling approaches in *NURBS Map I / II* or coding techniques in *Kyu Furukawa Teien Shūsaku*, each has been used in order to produce a rigorous and objective foundation so that myriad creative layers of explorations can unfold from this base. Thus, rather than relying purely on the self-referential nature of abstract mathematics—that is, that the language of these results remains trapped within the confines of formulaic constructs—I have sought to usurp them in a practical manner that, at its end point, becomes a pursuance towards the *applied* rather than the described.

This then approaches what Julian Klein contends is the basis for 'artistic research.' Here, and in terms of my own investigations, Klein argues that the disciplines of art and science are in fact inevitably linked through their desire to produce new knowledge, and what separates them is not the context or framework of discovery, but simply the means and approaches each embraces to achieve those goals:

"Art and science are not separate domains, but rather two dimensions in the common cultural space. This means that something can be more or less artistic, while nothing would be already said about the amount of it being scientific. This is also true for many other cultural attributes, such as the musical, philosophical, religious or mathematical. Some of them are, on the contrary, more dependent on each other than they are isolated."[5]

5 Julian Klein, "What is Artistic Research," *Gegenworte* 23 (2010): 25.

But these interrelationships between disciplines or cultural attributes also means that for Klein, a distinction must be made between the subjectivity of art and artistic practice and the objectivity of research and science. Thus Klein acknowledges that though not all art is research, research can become an artistic pursuit:

"If 'art' is but a mode of perception, then also 'artistic research' must be the mode of a process. Therefore, there can be no categorical distinction between 'scientific' and 'artistic' research - because the attributes independently modulate a common carrier, namely, the aim for knowledge within research. Artistic research can therefore always also be scientific research [...] Against this background the phrase 'art as research' seems to be not quite accurate, because it is not the art, which evolves into research somehow. What exists, however, is research that becomes artistic – so it should be rather named 'Research as Art,' with the central question: When is Research Art?"[6]

For the concept of *spatial thinking*, the notion that research may be defined within an artistic framework, or at least a framework that enables objects or artifacts to become produced via a process that continually questions where current knowledge resides, allows for a greater flexibility regarding the utility of the objects and processes that are being made and refined. That these products might be equally positioned as both artistic objects as well as rigorous methodologies for spatial analysis, indeed allows for a further argument for Klein's observation that art and science are but two dimensions within a common cultural space. But given that the catalyst for this book has been the Japanese garden, and its obvious exemplar spatial proclivities regarding landscape and soundscape design, the framework of 'research as art' is one that has been driven here through a quest to not only describe, but then also transmediate the garden as both an artistic artifact as well as a rigorous series of spatial relationships.

As a starting point then for this book is the observation that Japanese garden design is an embodiment of an artistic process in which nature itself is *crafted*. The Japanese garden is certainly a crafted space, a space of considered balance, and a space of the meeting between human creativity and nature. Indeed it is through simple observation and deduction that one can immediately ascertain that though this crafting is a masterful imitation of nature, it is also a vehicle to allow for the role of the garden maker and garden designer to remain in close proximity to its source of imitation. Thus a gardener, much like the role that Kepler saw for himself, is as one who *reveals* nature, draws our eyes and ears to a beauty that was always there, or in the least, frames the aesthetics of nature as occupying an apogee that human endeavor should itself strive for. This then is exactly what Cage sought to create by using musical frames and compositional structures as a conduit through which artistic experiments in the unknown could emerge. But for Cage, these experiments were also vehicles to embrace not only the character of Zen, but also, and more distinctly, the

6 Klein, "What is Artistic Research," 26.

powerful currency that natural consequences and natural processes hold for generating an artifact free of a composer's premeditated ideas, values or intentions.

But the dialectic regarding the Japanese garden as a function of human artistic *production*, yet created to seem entirely natural, similarly plays against what Martin Heidegger saw as the relationship between the ancient Greek notion of *technê* (τέχνη) and *poïesis* (ποιέω). For Heidegger, *technê* was "the name not only for the activities and skills of the craftsman, but also for the arts of the mind and the fine arts."[7] Though as Mahon O'Brien has pointed out, Heidegger's interpretation of *technê* is in the narrow "elemental/original sense of technê as suggested by Aristotle's discussion of the difference between technê and pratakton." Whereas Aristotle considered *pratakton* to refer to the practical or perfunctory matters of making or production, *technê* is a means in which a conception of the making as a *manufacturing*, as well as an aesthetic positioning is preconceived. Heidegger's interpretation then of the ancient Greek notion of *poïesis* as a "bringing forth," or a type of blooming or transition resides in the process of *technê* in that it is a means to reveal the real essence of things. For the artistic and analytical work I have presented in this book, a shared goal of both identifying and then revealing the essence of the Japanese garden has been facilitated through a similar path as that that Heidegger suggested. This approach then is essentially aligned with the phenomenological notion that technical making is a way of realizing meaning in an artifact. As Andrew Freenberg explains:

"The idea of the artifact is not arbitrary or subjective but rather belongs to a technê. Each technê contains the essence of the thing to be made prior to the act of making. The idea, the essence of the thing is thus a reality independent of the thing itself and its maker. What is more [. . .] the purpose of the thing made is included in its idea."[8]

Throughout my own investigations and development of the approach of *spatial thinking*, a primary concern has also been to forge a closer nexus between what can be described as the state of theoretical knowledge (or as in Greek, *epistemê)* to the practice of making and crafting (that of *technê*). These two areas of interest have also had a long history in terms of their ancient Greek roots and modern-day re-interpretations and usurpations within myriads fields such as art, architecture and design. But as Richard Parry notes, the duality between the notions of theory and practice, at least according to ancient Greek philosophy, is one of intimacy and even mutual necessity:

"*Epistemê* is the Greek word most often translated as knowledge, while *technê* is translated as either craft or art. These translations, however, may inappropriately

7 Martin Heidegger, *The Question Concerning Technology and other essays*, trans. William Lovitt (New York: Harper & Row, 1977), 13.
8 Andrew Freenberg, *Heidegger and Marcuse: the catastrophe and redemption of history* (New York: Routledge, 2004), 1.

harbor some of our contemporary assumptions about the relation between theory (the domain of 'knowledge') and practice (the concern of 'craft' or 'art'). Outside of modern science, there is sometimes skepticism about the relevance of theory to practice because it is thought that theory is conducted at so great a remove from reality, the province of practice, that it can lose touch with it [. . .] Within science, theory strives for a value-free view of reality. As a consequence, scientific theory cannot tell us how things should be — the realm of 'art' or 'craft' [. . .] The relation, then, between *epistemê* and *technê* in ancient philosophy offers an interesting contrast with our own notions about theory (pure knowledge) and (experience-based) practice. There is an intimate positive relationship between *epistemê* and *technê*, as well as a fundamental contrast."[9]

I position *spatial thinking* then as an antidote to the modern distraction that has placed theory and practice as not within easy reach of each other, but in an almost defiant opposition. This bifurcation has been suggested by Paul Carter as one way in which the discipline of acoustic ecology has found its initially strong 1970s activism and environmentalism quelled under the fixation by new soundscape composers of the digital tools for spatial audio recording and the processes of musical production and acoustic simulation. But it is not only acoustic ecology that has been seduced or sidetracked by an emphasis towards reaching either for a new *epistemê* or *technê* rather than seeking a nexus to conjoin them. Shane Murray sees the "high theory" years of 1980s in the field of architecture, with its pursuit of Jacques Derrida's theory of deconstruction, as well as the notions of Marcus Novak and other digital provocateurs regarding the possibilities of the datascape, as a dead end, and furthermore, as a position that has willingly disconnected itself from research into the practice of architectural design as an act of making.[10] Perhaps for the discipline of architecture, this engagement with new technologies and thus the greater need for ways and means in which to conceptualize and theorize these technological avenues has created a more visible divide between understanding knowledge as a *pre-design* act compared to knowledge *as* the act of design.

But for the framework I have developed around *spatial thinking*, it is the importance of the *epistemê* as a way in which to inform the direction of the *technê*. Both are necessary in order for the work of art to find an end point, and inevitably both are a function of technological tools that enable new directions or paths to unfold. These technological tools and techniques have of course varied and been variably applied and re-applied within all of the projects described here, such that they complement what Agostino Di Scipio's remarks as the continued usefulness of conceptualizing

9 Richard Parry, "Episteme and Techne," in *The Stanford Encyclopedia of Philosophy*, ed. Edward N. Zalta, accessed September 24, 2013, http://plato.stanford.edu/archives/fall2008/entries/episteme-techne/.

10 Shane Murray, "Architectural Design and Discourse," *Architectural Design Research* 1/1 (2005): 83-102.

technê. For Di Scipio *technê* is a means for which "art [is] made by inventing the techniques of its making, which is to say by questioning established, inherited techniques and methods."[11] This has been true of the work discussed in this book in which methodologies such as Voronoi tessellation, Greimas Square, CA, FCA and NURBS modeling are taken out of the contexts from which they are normally embedded, in order to push the boundaries of how Japanese garden design can be conceptualized such that new modes of spatial thinking begin to flourish.

As a theoretical construct though, such a usurpation of new technologies as drivers of artistic exploration is also an example of what Theodor Adorno considers as the *Technifizierung* (technification) of the artwork. On describing the impact within music composition, Adorno comments that:

"The technification of the work of art matures along with the inclusion of techniques that had developed outside music in the course of the general growth of technology [. . .] The concept of musical technique grew in complexity in a way comparable to the development of film. Means that did not flow from composition in the first instance were adopted and helped to enlarge its scope."[12]

Though Adorno became critical of the ratio for which the technology might impact on the aesthetics of a work, he saw the uptake of new technologies into artistic practice as an inevitability of the modern world. Indeed, for Adorno, if art was to be truly modern, it should seek to use the subject's experience of technology and expose the tension between the aesthetic and technological forces at play. Martin Dixon elaborates:

"By incorporating both technological means and the subject's experience of technology within its own productive procedures - acknowledging respectively a 'neutral' stance regarding technology and its ideological hegemony - art becomes a means of diverting the remorseless momentum of modern technology for the purposes of its own inner-aesthetic organization."[13]

Adorno's views on the role of technology in society are akin to the technologically determinist philosophy espoused by Raymond Kurzweil[14] and explained by M. R. Smith

11 Agostino Di Scipio, "Towards a critical theory of (music) technology: computer music and subversive rationalization," in *International Computer Music Conference Proceedings* (Thessaloniki, Greece: ICMC, 1997), 62.

12 Theodor Adorno, *Sound Figures*, trans. R. Livingston (Standford, CA: Stanford University Press, 1999), 199.

13 Martin Dixon, "Adorno on Technology and the Work of Art," *ARiADA* 1/1 (2000): 1-18.

14 Raymond Kurzweil, *The Singularity is Near* (London: Gerald Duckworth & Co. Ltd, 2005).

as a belief that "changes in technology exert a greater influence on societies and their processes than any other factor."[15]

In regards to the Japanese garden though and the concept of *spatial thinking*, technology and technological tools have certainly enabled new forms of interdisciplinary investigations to arise, though these interrogations—of the spatial information of the garden and the concept that the garden represents a type of exemplar—are a function not of the novelty of the tools themselves, but the potential they posses for generating innovative methodologies that reveal new knowledge about the gardens. *Spatial thinking* is then a method for thinking *through* an object. This is perhaps an invitation for re-reading the work and methodologies of Sesshū, whose grasp of highly abstract concepts of space and spatial relationships became transferred not only within the two-dimensional plane of the rice-paper scroll, but the three-dimensional space of the temple grounds. That Sesshū explored the composition of landscapes and landscape forms through various media implies that he was able to think through fundamental structures and then choose the most aesthetically appropriate or readily available media that would embody these structures. Of course for Sesshū these structures were identified, like Pythagoras and Kepler before him, within phenomena that were observed within extant natural world. That the concept of nature and the natural environment acted as a paradigm for Sesshū, in which his compositions were created in both ink as well as with rocks and earth, stemmed not only from his Buddhist beliefs but his travels through Ming-era China. His traveling companion, the Buddhist poet Genryū, noted of their journey in China that:

"Sesshū said, 'Within the great country of China, there are no painting masters. That is not to say that there are no paintings, but no painting teachers there except as there are mountains [. . .] and as there are rivers [. . .] strange plants, trees, birds, and beasts, different men and their manners and customs. Such are the real paintings of China. In regard to the techniques of using ink and the art of handling the brush, these must be mastered first in the mind and then translated by the hand."[16]

Sesshū's art of ink painting and garden making was then a means in which to think through those structures, materialities and forms evident in the landscapes of Japan and China. Because of this approach he developed in painting what Dore Ashton notes as "a new vocabulary of signs—angular, broadly brushed planes that could establish vast spaces with a minimum of detail. He originated many inventions to modulate space, and these inventions could be considered at once descriptive and abstract."[17] This interaction between the abstract and the descriptive or the purely

15 Michael L. Smith, "Technological Determinism in American Culture," in *Does Technology Drive History: the dilemma of technological determinism*, eds. Merritt Roe Smith and Leo Marx (Boston: MIT Press, 1994), 2.

16 Jon Carter Covell, *Under the Seal of Sesshū*, (New York: Hacker Art Books, 1975), 24.

17 Dore Ashton, *Noguchi East and West* (Berkley: University of California Press, 1992), 98.

representational is also evident at the Sesshutei garden at Joei-ji in which those most basic elements of an idealized landscape such as the presence of famous mountain peaks like Mt. Fuji, *o-karikomi* and the *karetaki*, are offset through a manipulation and flattening of the picture space from the seated viewing area of the veranda. Those principle elements of the natural world were used both in representational terms as well as manipulated in their relationships to one another. Sesshū then was able to identify structural traits of nature, but in a much less formal manner than Kepler or Pythagoras, he manifested these proclivities into art works that both expressed themselves as images *after* nature and objects *of* nature.

Spatial thinking then seeks in a similar vein to explore alternative embodiments of a set of core fundamental principles or proclivities expressed in a number of media. It is enacted then through myriad methodologies that seek to interrogate, assess, re-form and understand an object or paradigm, though is not reliant on purely closed disciplinary excursions but rather seeks to open up the object of inquiry for inter-disciplinary discussion. Thus if Sesshū's paradigm was nature itself, then the paradigm I have followed is that of the Japanese garden. As such, *spatial thinking* is also useful as an analytical method that examines not simply the surface features—the obvious traits or characteristics—but moreover, those deeper relationships that remain unseen within an object. It is thus about generating an *epistemê* using formal approaches with the help of technologies, mathematical abstractions and technical apparatuses in order to use these to guide methods for developing myriad *technê*. In this respect, such creative outcomes partake in the larger methodological framework and are not separable as singular entities that are self-referential or a-contextual. It can be perhaps equally equated to how Simon Blackburn describes the critical theory of structuralism as "the belief that phenomena of human life are not intelligible except through their interrelations. These relations constitute a structure, and behind local variations in the surface phenomena there are constant laws of abstract culture."[18] If the Japanese garden can be read as a series of parts that constitute a whole, and if the relationships between the parts are privy to particular rules, laws or aesthetic associations, then the goal of *spatial thinking* is to identify these constituents. Its role is to think through what binds them together, their materialities and forms, and then to reconstruct these bonds of material flows into secondary mediums such that the garden itself is traversed, yet at the same type becomes an agent that traverses other future spaces, realms and media.

18 Simon Blackburn, *Oxford Dictionary of Philosophy* (Oxford: Oxford University Press, 2008), 353.

Bibliography

Adams, Mags, Trevor Cox, Gemma Moore, Ben Croxford, Mohamed Refaee, and Steve Sharples. "Sustainable Soundscapes: Noise Policy and the Urban Experience." *Urban Studies* 43, no. 13 (December 2006): 2385-2398.

Adorno, Theodor. *Sound Figures.* Stanford, CA: Stanford University Press, 1997.

Adorno, Theodor. "Vers une musique informelle." In *Quasi una fantasia: Essays on Modern Music*, by Theodor Adorno, translated by Rodney Livingstone, 269-322. London: Verso, 1998.

Alison, Jane, Marie-Ange Brayer, and Frédéric Migayrou, editors. *Future City.* London: Thames and Hudson, 2006.

Almenberg, Gustaf. *Notes on Participatory Art: Towards a Manifesto Differentiating it from Open Work, Interactive Art and Relational Art.* Central Milton Keynes: AuthorHouse, 2010.

Aoki, Yoji, Hitoshi Fujita, and Koichiro Aoki. "Measurement and analysis of congestion at the traditional Japanese garden 'Korakuen'." In *Monitoring and Management of Visitor Flows in Recreational and Unprotected Areas Conference Proceedings*, edited by A. Arnberger, C. Brandenburg and A, Muhar, 264-270. Vienna: Bodenkultur University, 2002.

Arkette, Sophie. "Sounds like City." *Theory, Culture & Society* 21 (2004): 159-168.

Ashton, Dore. *Noguchi, East and West.* Berkley: University of California, 1992.

Augoyard, Jean François, and Henry Torgue. *Sonic experience: a guide to everyday sounds.* Translated by Andrea McCarthy and David Paquette. Montreal: McGill-Queen's University Press, 2005.

Aurenhammer, F. "Voronoi Diagrams-A Survey of a Fundamental Geometric Data Structure." *ACM Computing Surveys* 23, no. 3 (1991): 345-404.

Auvreli, P. V. *The Project of Autonomy: politics and architecture within and against capitalism.* New York: Princeton Architectural Press, 2008.

Bachnik, Jane M. "Orchestrated reciprocity: Belief versus practice in Japanese funeral ritual." In *Ceremony and Ritual in Japan, religious practices in an industrialized society*, edited by Jan van Bremen and D. P. Martinez, 108-128. New York, NY: Routledge, 2003.

Bandt, Ros. *Sound Sculpture, Intersections in Sound and Sculpture in Australian Artworks.* Sydney: Fine Arts Press, 2001.

Barker, Roger. *Ecological Psychology.* Standford: Stanford University Press, 1968.

Bays, C. "Introduction to Cellular Automata and Conway's Games of Life." In *Game of Life Cellular Automata*, edited by A. Adamatzky, 1-7. London: Springer-Verlag, 2010.

Beck, Robert. "Spatial Meaning and the Properties of the Environment." In *Environmental Perception and Behavior*, edited by D. Lowenthal, 18-41. Chicago: University of Chicago, 1967.

Bernier, Bernard. *Breaking the cosmic cycle: religion in a Japanese village.* Ithaca, NY: Cornell University, 1975.

Berthier, François. *Reading Zen in the rocks: the Japanese dry landscape garden.* Chicago, IL: University of Chicago Press, 2000.
Blackburn, Simon. *Oxford Dictionary of Philosophy.* Oxford: Oxford University Press, 2008.
Blesser, Barry, and Ruth Linda Salter. *Spaces speak, are you listening: experiencing aural architecture.* Boston: MIT Press, 2007.
Bognar, Botand. "Surface above all? American influence on Japanese urban space." In *Transctions, transgessions, transformations: American culture in Western Europe and Japan,* edited by Heide Fehrenbach and Uta. G. Poiger, 45-80. New York: Berghahn Books, 2000.
Bognar, Botond. "Anywhere in Japan." *Architecture Design* 62, no. 11-12 (1992): 70-72.
Bossy, Michel-André, Thomas Brothers, and John C, McEnroe. *Artists, Writers and Musicians.* Westport: Oryx Press, 2001.
Branzi, A. *No-Stop City: Archizoom Associati.* Orleans: Hyx, 2006.
Bring, Mitchell, and Josse Wayembergh. *Japanese Gardens: Design and Meaning.* New York: McGraw-Hill, 1981.
Broadbent, Jeffrey. "Recent developments in architectural semiotics." *Semiotica* 101, no. 1-2 (1994): 73-101.
Bruhn, Siglind. *The Musical Order of the World: Kepler, Hesse, Hindemith.* Hillsdale, New York: Pendragon Press, 2005.
Burmeister, Peter, and Richard Holzer. "Treating Incomplete Knolwedge in Formal Concept Analysis." In *Formal Concept Analysis: Foundations, Applications,* 114-127. Berlin: Springer-Verlag, 2005.
Burrow, Andrew, and Robert F. Woodbury. "Whither Design Space?" *Artificial intelligence for Engineering Design, Analysis and Manufacturing* 20 (2006): 63-82.
Cage, John. *For the Birds.* Hannover, NH: Weslyan University Press, 1995.
—. "Ryoanji for oboe and obbligato percussion." New York, NY: Henmar Press, 1983.
Cage, John. "Ryoanji: Solos for Oboe, Flute, Contrabass, Voice, Trombone with Percussion or Orchestral Obbligato." *A Journal of Performance Art* (MIT Press) 31, no. 3 (September 2009): 57-64.
—. "*Ryoanji*: Solos for Oboe, Flute, Contrabass, Voice, Trombone With Percussion or Orchestral Obbligato (1983–85)." *A Journal of Performance and Art* 31/3 (2009): 57–64.
—. *Silence.* Middletown, CT: Wesleyan University Press, 1961.
—. "Variations IV." New York, NY: Henmar Press, 1963.
Carlson, Allen. *Aesthetics and the Environment, the appreciation of nature, art and architecture.* London: Roughtlege, 2000.
Carlyle, Angus. "Like Quails Clucking: Landscape, Soundscape and a Noisy Future." In *The East Meets West in Acoustic Ecology,* edited by Tadahiko Imada, Kozo Hiramatsu and Keiko Torigoe, 109-118. Hirosaki: JASE & HIMC, 2006.
Carrol, Nöel. "Cage and Philosophy." *The Journal of Aesthetics and Art Critisim* 52, no. 1 (Winter 1994): 93-98.

Carter, Paul. "Auditing Acoustic Ecology." *Soundscape, the journal of acoustic ecology* 4, no. 2 (2003): 12-13.

Casalis, Matthieu. "The semiotics of the visible in Japanese rock gardens." *Semiotica* 44, no. 3-4 (1983): 358-359.

Chiu, Che Bing, trans. *Yuanje le traite du jardin (1634).* New York, NY: Besacon, 1997.

Clark, C., and S. A. Stansfeld. "The Effect of Transportation Nosie on Health and Cognitive Development: A Review of Recent Evidence." *International Journal of Comparative Psychology* 20, no. 2 (2007): 145-158.

Conder, Josiah. *Landscape Gardneing in Japan.* New York: Dover, 1964.

Cook, Haruko Minegishi. *Socializing Identities Through Speech Style: Learners of Japanese as a Foreign Language.* New York, NY: Multilingual Matters Ltd., 2008.

Daniel, Terry, and Joanne Vinning. "Assessment of Landscape Quality." In *Behaviour and the Natural Environment*, edited by Irwin Altman and Joachim F. Wohlwill, 39-84. New York: Plenum Press, 1983.

Davey, B. A., and H. A. Priestly. *Introduction to Lattices and Order.* Cambridge: Camridge University Press, 2002.

Davy, Kyle V. "The Design Process, catching the third wave." In *Architecture celebrating the past, designing the future*, edited by Nancy B. Solomon, 158-164. New York: Visual Reference Publications, 2008.

de Barry, William Theodore. *Sources of Japanese Tradition.* 2nd Edition. Edited by William Theodore de Barry. Vol. I. New York: Columbia University Press, 2001.

Delio, Thomas. "John Cage's *Variations II*: Morphology of a Global Structure." *Perspectives of New Music* 19, no. 1-2 (1981): 351-371.

Dennett, D. C. *Darwin's Dangerous Idea: Evolution and the Meanings of Life.* New York: Simon & Schuster, 1995.

Di Scipio, Agostino. "Towards a critical theory of (music) technology: computer music and subversive rationalization." *International Computer Music Conference Proceedings.* Thessaloniki: ICMC, 1997. 62-65.

Dreyer, C. "The crisis of representation in contemporary architecture." *Semiotica* 143 (2003): 163-183.

Evans, Robin. *Translations from Drawing to Building and Other Essays.* London: Architecture Association, 1997.

Faulkner, Rupert. *Masterpieces of Japanese Prints.* Tokyo: Kodansha International, 1991.

Fontanille, J. *Sémiotique du discours.* Limoges: Presses de l'Université de Limoges, 2003.

Forte, Allen. *The structure of atonal music.* New Haven: Yale Univerity Press, 1973.

Foucault, Michel. "Of Other Spaces, Heterotopias." *Architecture, Mouvement, Continuité* 5 (1984): 46-49.

Fowler, Michael. "Hearing a shakkei: the semiotics of the audible in a Japanese stroll garden." *Semiotica* 197 (October 2013): 101-117.

—. "Sound, aurality and critical listening: disruptions at the boundaries of architecture." *Architecture and Culture* 1/1 (November 2013): 159-178.

—. "The taxonomy of a Japanese stroll garden: an ontological investigation using Formal Concept Analysis." *Axiomathes* 23, no. 1 (March 2013): 43-59.

—. "Towards an urban soundscaping." In *Infrastrukturen des urbanen: Soundscapes, Netscapes, Landscapes im kinematischen Raum*, edited by Nathalie Bredella and Chris Dähne, 35-57. Bielefeld: Transcript, 2013.

—. "Appropriating an architectural design tool for musical ends." *Digital Creativity* 22/4 (January 2012): 275-287.

—. "Transmediating a Japanese garden through spatial sound design." *Leonardo Music Journal* 21 (December 2011): 41-49.

—. "Mapping sound-space: the Japanese garden as auditory model," *Architectural Research Quarterly* 14, no. 1 (March 2010): 63-70.

—. "Shin-Ryoanji: a digital garden." *Soundscape, The journal of Acoustic Ecology* 8/1 (Fall/Winter 2008): 33-36.

Frank, Lee. "Media as input to Perceptual Systems." *www.leefrank.com.* 1969. http://www.leefrank.com/inquiry/mind/consciousness/media.html (accessed March 18, 2013).

Franklin, Ursula. "Silence and the notion of commons." *Soundscape, the journal of acoustic ecology* 1, no. 2 (Winter 2000): 14-17.

Freenberg, Andrew. *Heidegger and Marcuse: the catastrophe and redemption of history.* New York: Routledge, 2004.

Furukawa, Kiyoshi, Haruyuki Fujii, and Yashuhiro Kiyomizu. "A comparative study on the relationship between music and space experience, based on tempo at a circuit style garden." *IPSJ SIG Technical Reports* 19 (2006): 7-12.

Ganter, B., G. Stumme, and R. Willie, eds. *Formal Concept Analysis; foundations and applications.* Berlin: Springer Verlag, 2005.

Gardner, M. "Mathematical Games: The fantastic combinations of John Coway's new solitaire game 'Life'." *Scientific American* 233 (1970): 120-131.

Gehring, Katrin, and Ryo Kohsaka. "'Landscape' in the Japanese language: Conceptual differences and implications for landscape research." *Landscape Research* (Taylor and Francis Group) 32, no. 2 (April 2007): 273-283.

Gengnagel, C., Kilian, A., Palz, N., Scheurer, F., eds. *Computational Design Modeling - Proceedings of Design Modeling Symposium Berlin.* Berlin: Springer-Verlag, 2011.

Gerimas, A. J. *Sémantique structurale.* Paris: Larousse, 1966.

Gibson, James. *The Senses Considered as Perceptual Systems.* Boston: Houghton Mifflin, 1966.

Gidlöf-Gunnarsson, A., and E. Öhrström. "Noise and well-being in urban residential environments: The potential role of perceived availability to nearby green areas." *Landscape and Urban Planning* (Elsevier) 83, no. 2-3 (2007): 115-126.

Greimas, Algirdas. *Structural Semantics: An Attempt at a Method.* Translated by Daniele McDowell, Ronald Schleifer and Alan Velie. Lincoln: University of Nebraska Press, 1983.

Greimas, Algirdas. "The Interaction of Semiotic Constraints." *Yale French Studies* 41 (1968): 86-105.

Greimas, Algirdas, and Jean Fontaille. *The Semiotics of the Passions: from states of affairs to states of feelings*. Translated by P. Perron and F. Collins. 1991.
Guastavino, Catherine. "The Ideal Urban Soundscape: Investigating the Sound Quality of French Cities." *Acta Acustica* 92 (2006): 945-951.
Guédon, Jean-Claude. "Architecture as Transdisicplinary Knowledge." In *Anyplace*, edited by Cynthia C. Davidson, 88-99. Cambridge: MIT Press, 1995.
Gulthorpe, Mark. *The Possibility of an Architecture*. New York: Routledge, 2008.
Guth, Christine M. E. *Art, Tea, Industry, Masuda Takashi and the Mitsui Circle*. Princeton: Princeton University Press, 1993.
Guy, Richard, K. "Conway's Prime Producing Machine." *Mathematics Magazine* 56, no. 1 (January 1983): 26-33.
Hane, Mikiso. *Modern Japan: A Historical Survey*. Boulder: Westview Press, 2001.
Harada, Jiro. *Gardens of Japan*. New York: Routledge, 2009.
Harvey, Lawrence. "Melbourne's Urban Electroacoustic Soundscape Systems.." Paper 2010. http://www.rmit.edu.au/architecturedesign/sial/soundstudio/projects/urbansoundscape (acessed January 24, 2014).
Hattori, Tsutomu. "Garden Formation of Kosihikawa-Korakuen Garden based on 'The Plan of Mitosama Koishikawa Oyashihi's Garden'." *Journal of Agricultural Science* 44, no. 1 (1999): 19-29.
Hays, K. Michael. *Architecture Theory since 1968*. Cambridge: MIT Press, 2000.
—. *Architecture's Desire: Reading the Late Avant-Garde (Writing Architecture)*. Cambridge: MIT Press, 2010.
Hébert, Louis. *Dispositifs pour l'analyse des textes et des images*. Limoges: Presses de l'Université de Limoges, 2007.
Heford, Per, and Per G. Berg. "The Sounds of Two Landscape Settings: Auditory concepts for physical planning and design." *Landscape Research* 28, no. 3 (2003): 245-263.
Heidegger, Martin. *The Question Concerning Technology and other essays*. Translated by William Lovitt. New York: Harper & Row, 1977.
Hellström, Bjorn, M. E. Nilsson, and P. Becker. "Acoustic design artifacts and methods for urban soundscapes." *Proceedings of the ICSV'08 (5th International Congress on Sound and Vibration)*. Daejeon: ICSV, 2008.
Helms, Hans G. *John Cage talking on music and politics*. Cassette tape. München: Edition S-Press, 1975.
Holborn, Mark. *The Ocean in the Sand*. London: Gordon Fraser, 1978.
Holl, Steven. "Stretto House." In *Architecture as a translation of music*, edited by Elizabeth E. Martin, 56-59. New York: Princeton Architectural Press, 1994.
Hoover, Thomas. *Zen Culture*. London: Arkana, 1977.
Iiachinski, Andrew. *Cellular Automata, a discrete universe*. London: World Scientific, 2001.
Imada, Tadahiko. "Acoustic Ecology Considered as a Connotation: Semiotic, Post-Colonial and Educational Views of Soundscape." *Soundscape, the journal of acoustic ecology* 11, no. 2 (2005): 13-17.

Ingold, Tim. "Against Soundscape." In *Autumn Leaves: sound and the environment in artist practice*, edited by Angus Carlyle, 10-13. Paris: Double-Entendre, 2009.
Itoh, Teiji. *Space and Illustion in the Japanese Garden*. New York, NY: Weatherhill, 1983.
—. *The Japanese Garden: an Approach to Nature*. New Haven, CT: Yale University Press, 1972.
Izutsu, Toshihiko. "Tokyo-Montreal: The Elimination of Color in Far-Eastern Art and Philosophy." In *The Realms of Color*, edited by Adolf Portmann and Rudolf Ritsema, 428-464. Dallas: Spring Publications, 1974.
Jen, Erica. "Aperiodicity in One-dimentional Cellular Automata." In *Cellular Automata: Theory and Experiment*, 3-18. Amsterdam: Elsevier Science Publishers, 1990.
Jonas, Marieluise. "Oku: the notion of the interior in Tokyo's urban landscape." In *Urban Interior*, edited by Rochus Hinkel, 99-112. Baunach: Spurbuchverlag, 2011.
Joseph, Brandon W. "John Cage and the Architecture of Silence." *October* (MIT Press) 81 (Summer 1997): 80-104.
Kōshirō, Haga. "The Wabi Aesthetic through the Ages." In *Essays on the History of Chanoyu*, edited by Paul Varley and Isao Kumakura, translated by Martin Collcuttin, 195-232. Honolulu: Unievrsity of Hawaii Press, 1989.
Kang, Jian. "From understanding to designing soundscapes." *Frontiers of Architecture and Civil Engineering in China* 4, no. 4 (2010): 403-417.
Kawarada, Mieko, and Taiichi Itoh. "Environmentalism in Japanese Gardens." In *Environmentalism in Landscape Architecture*, edited by Michael Conan, 245-268. Washington, D.C: Dumbarton Oaks, 2000.
Keane, Marc Peter. *Japanese garden design*. Tokyo: Tuttle Co., Inc., 1996.
Keller, Damián, Ariadna Capasso, and Scott Wilson. "Urban Corridor: Accumulation and Interaction as Form-Bearing Process." *International Computer Music Conference Proceedings*. Göthenburg: ICMC, 2002. 295-298.
Kepler, Johannes. *The Harmony of the World*. Translated by E. J. Aiton, A. M. Duncan and J. V. Field. Philadelphia: American Philosophical Society, 1997.
Kihara, Toshie. "The Search for Profound Delicacy: The Art of Kano Tan'yu." In *The Arts of Japan: An international Symposium*, edited by Miyeko Murase and Judith G. Smith, 248-251. New York: Metropolitan Museum of Art, 2000.
Kinoe, Yosuke, and Hirohiko Mori. "Mutual Harmony and Temporal Continuity: A Perspective from the Japanese Garden." *ACM SIGCHI Bulletin* 25, no. 1 (1993): 10-13.
Kipnis, J. "On Criticism." *Harvard Design Magazine* Fall (2005): 96-104.
Klein, Julian. "What is Artistic Research. *Genegenworte* 23 (2010): 25-29.
Kostelanetz, Richard. *The Theatre of Mixed Means*. London: Pitman, 1970.
—. *John Cage (ex)plain(ed)*. New York: Schirmer Books, 1996.
—. *Conversing with Cage*. London: Routledge, 2003.
Krause, Bernie. "Anatomy of a Soundscape: Evolving Perspectives." *Journal of the Audio Engineering Society* 56, no. 1-2 (2008): 73-101.

Kubke, M. Fabiana, and Catherine E. Carr. "Development of Auditory Centers Responsible for Sound Localization." In *Sound source localization*, edited by Arthur N. Popper and Richard R. Fray, 179-237. New York: Springer, 2005.

Kuck, Lorraine. *The world of the Japanese Garden*. New York, NY: Weatherhill, 1968.

Kuitert, Wybe. *Themes in the history of Japanese garden art*. Honolulu, HI: University of Hawai'i Press, 2002.

Kurzweil, Raymond. *The Singularity is Near*. London: Gerald Duckworth & Co. Ltd., 2005.

Lefebvre, Henri. *The Production of Space*. Translated by Donald Nicols-Smith. Oxford: Blackwell Publishing, 1974.

Leus, Maria. "The soundscape of cities: a new layer in city renewal." In *Sustainable Development and Planning V*, edited by C. A. Brebbia and E. Beriatos, 355-367. Southhampton: WIT Press, 2011.

Lindverg-Wada, Gunilla. *Japanese Literary History: The Beginings*. Vol 1, in *Literary History: Towards a Global Perspective*, 111-134. Berlin: Walter de Gruyter, 2006.

Litton, R. B. "Aesthetic Dimensions of a Landscape." In *Natural Environments: Studies in Theoretical and Applied Analysis*, 284-286. Baltimore: John Hopkins University Press, 1972.

Lo, Yi-Luen, and Mark D. Gross. "Thinking with Diagrams in Architectural Design." *Artificial Intelligence Review* 5 (2001): 135-149.

Lordick, Daniel. *Curve Surface Freeforms: Rhinoceros for Architects*. New York: Springer-Verlag, 2009.

Lynch, Kevin. *City Sense and City Design: Writings and Projects of Kevin Lynch*. Edited by Tridib Banerjee and Michael Southworth. Boston, MA: MIT Press, 1995.

Lynn, Greg. *Animate Form*. Princeton: Princeton Architectural Press, 1999.

Lyon, Eric. "Spectural Tuning." *International Computer Music Conference Proceedings*. Miami: ICMC, 2004. 375-377.

Main, Alison, and Newell Platten. *The Lure of the Japanese Garden*. Kent Town, South Australia: Wakefield Press, 2005.

Manolopoulou, Yeoryia. "The Active Voice of Architecture: An Introduction to the Idea of Chance." *Field* 1 (2007): 62-72.

Martinez, D. P. "When Soto becomes Uchi: Some thoughts on the Anthropology of Japan." In *Dismantling the East-West Dichotomy*, edited by Joy Hendry and Heung Wah Wong, 31-37. New York, NY: Routledge, 2006.

McClary, Susan. *Feminine Endings: Music, Gender and Sexuality*. Minneapolis: University of Minnesota Press, 1991.

Migayrou, Frédéric. "non-standard order: 'nsa codes'." In *Future City*, edited by Jane Alison, Marie-Ange Brayer and Frédéric Migayrou, 16-33. London: Thames and Hudson, 2006.

Miller, Mara. *The Garden as an art*. New York, NY: State University of New York Press, 1993.

Missingham, Greg. "Japan 10±, China 1: A First Attempt at Explaining the Numerical Discrepancy between Japanese-style Gardens Outside Japan and Chinese-style Gardens Outside China." *Landscape Research* 32 (2007): 117-146.

Mitchell, W. J. *The Logic of Architecture: Design, Computation and Cognition.* Cmabridge: MIT Press, 1990.

Miura, Kayo, and Haru Sukemiya. "Visual impression of Japanese rock garden (karesansui): from the point of view of spatial structure and perspective cues." *Proceedings of the International symposium on ecoTopia science.* Nagoya: ISETS, 2007. 1165-1168.

Morag, J. Grant. *Serial music, serial aestehtics: composition theory in post-war Europe.* New York: Cambridge University Press, 2001.

Murase, Miyeko. *Bridge of Dreams, The Mary Griggs Burke Collection of Japanese Art.* New York: The Metropolitan Musuem of Art, 2000.

—. *L'Art du Japon.* Paris: La Pochothéque, 1996.

Murray, Shane. "Architectural Design and Discourse." *Architectural Design Research* 1, no. 1 (2005): 83-102.

Nakamura, Makoto. "The Twofold beauties of the Japanese garden." In *International Federation of Landscape Architects (I.F.L.A) Yearbook*, 195-198. Brussels: IFLA, 1986/87.

Naveh, Z., and A. S. Lieberman. *Landscape ecology: theory and application.* New York: Springer Verlag, 1984.

Naveh, Zeev, and Arthur S. Lieberman. *Landscape Ecology: Theory and Application.* New York: Springer-Verlag, 1984.

Nitschke, Günter. *Japanese Gardens, right angle and natural form.* Köln: Taschen, 1999.

Nouma, Seiroku. *The Art of Japan: Ancient and Medieval.* Tokyo: Kodansha International, 1966.

Novak, Marcus. "Computation and composition." In *Architecture as a translation of music*, edited by Elizabeth Martin, 66-69. New York: Princeton Architectural Press, 1994.

Novak, Marcus. "Liquid Architectures in Cyberspace." In *Cyberspace: First Steps*, edited by M. Benedikt, 225-254. Cambridge: MIT Press, 1991.

Nute, Kevin. *Place, time and being in Japanese architecture.* New York, NY: Routledge, 2003.

Ohno, Ryuzo, and Miki Kondo. "Measurement of the multi-sensory information for describing sequential experience in the environment: an application to the Japanese circuit style garden." *The Urban Experience: Proceedings of the 13th Conference of the International Association for People-Environment Studies.* Manchester: E & FN SPON , 1994. 425-437.

Ogawa, M., and K. L. Ma. "code_swarm: A Design Study in Organic Software Visualisation." *IEEE Transactions on visualization and computer graphics* 15, no. 6 (2009): 1097-1104.

Ostwald, Michael. "Architectural Theory Formulation through Appropriation." *Architectural Theory Review* 4, no. 2 (1999): 52-70.

Owen, William. "I saw a man he wasn't there." In *Mapping*, edited by William Owen, 154-155. Mies: RotoVision, 2005.

Oxman, R. "Performative design: a performance-based model of digital architectural design." *Environment and Planning B: Planning and Design* 26, no. 6 (2009): 1026-1037.

Pallasmaa, Juhani. *Eyes of the skin*. Chichester: Wiley-Academy, 2005.

Parisi, Luciana. *Contagious Architecture: Computation, Aesthetics and Space*. Cambridge: MIT Press, 2013.

Parker, Joseph D. *Zen Buddhist Landscape Arts of Early Muromachi Japan (1336-1573)*. New York: State University of New York Press, 1999.

Parkes, Graham. "Further Reflections on the Rock Garden of Ryōanji: From Yūgen to Kire-tsuzuki." In *The Aesthetic Turn: Reading Eliot Deutsch on comparative philosophy*, edited by Roger T. Ames, 13-28. Chicago, IL: Open Court Publishing Company, 1999.

Pérez-Gómez, Alberto. *Architecture and the Crisis of modern Science*. Cambridge: MIT Press, 1983.

Peterson, James. *Dreams of Chaos, Visions of Order: Understanding the American Avant-garde Cinema*. Detroit, IL: Wayne State University Press, 1994.

Potvin, D., K. M. Parris, and R. Mulder. "Geographically pervasive effects of urban noise on frequency and syllable rate of songs and calls in silvereyes (Zosterops lateralis)." *Royal Society of London Philosophical Transactions -Biological Sciences* 278, no. 1717 (2010): 2464-2469.

Pregil, Phillip, and Nancy Volkman. "Landscapes of the Rising sun: Design and Planning in Japan." In *Landscapes in History: Design and Planning in the Eastern and Western Traditions*, by Phillip Pregil and Nancy Volkman, 340-380. New York: John Wiley and Sons, 1999.

Priss, Uta. "Facet-like Structure in computer Science." *Axiomathes* 18 (2008): 243-255.

Priss, Uta. "Formal concept analysis in information science." *Annual review of information science and technology* 40 (2006): 521-543.

Pritchett, James. *The Music of John Cage*. Cambridge, MA: Cambridge University Press, 1993.

—. "David Tudor as Composer/Performer in Cage's Variations II." *Leonardo Music Journal* 14 (2004): 11-16.

Puckette, Miller. "Using Pd as a score language." *International Computer Music Conference Proceedings*. Gothenburg: ICMC, 2002. 184-187.

Puckette, Miller, Theodore Apel, and David Zicarelli. "Real-time Audio Analysis Tools for Pd and MaxMSP." *International Computer Music Conference Proceedings*. Ann Arbor: ICMC, 1998. 109-112.

Quinn, Charles. "Uchi/Soto: Tip of a semiotic iceberg? 'Inside' and 'outside' knowledge in the grammar of Japanese." In *Situated meaning: Inside and outside in Japanese self, society, and language*, edited by Jane M. Bachnik and Charles J. Quinn, 249-294. Princeton, NJ: Princeton University Press, 1994.

Raimbault, Manon, and Danièle Dubois. "Urban soundscapes: Experiences and knowledge." *Cities* 22, no. 5 (2005): 339-350.

Retallack, Joan. *Conversations in Retrospect.* Hannover, NH: University of New England Press, 1996.

Reynolds, C. "Flocks, herds and schools: A distributed behavioural model." *Computer Graphics* 21, no. 4 (1987): 289-296.

Ryan, Marie-Laure. "Diagramming narrative." *Semiotica* 2007, no. 165 (2007): 11-40.

Ryuzo, Ono, and Miki Kondo. "Measurement of the multi-sensory information for describing sequential experience in the environment: an application to the Japanese circuit garden." *A People-Environment Perspective: Proceedings 13th International Conference of the IAPS.* Manchester: E&FN Spon, 1994. 425-437.

Sacchi, Livio. *Tokyo, City and Architecture.* Torino: Skira Editore, 2004.

Sansom, George. *A History of Japan to 1334.* Tokyo: Charles E. Tuttle Co., 1963.

Schaeffer, Pierre. *Traité des objects musicaux.* Paris: Le Seuil, 1966.

Schafer, R. Murray. *The soundscape: our sonic environment and the tuning of the world.* Rochester, NY: Destiny Books, 1977.

Schumacher, Patrik. "Parametricism and the Autopoiesis of Architecture." *Log* 21 (2011): 63-79.

Seidensticker, E. G., trans. *The tale of Genji.* Vol. 1. Tokyo: Charles E. Tuttle Co., 1976.

Sherriden, Ted, and Karen van Lengen. "Hearing Architecture." *Journal of Architectural Education* 57, no. 2 (2003): 37-44.

Shigemori, Mirei. *Karesansui.* Kyoto: Kawara Shoten, 1965.

—. *Nihon teien-shi zukan.* Tokyo: Yukosha, 1936-39.

Shin, Hoon, Hyuk Song, Gi-Bong Nam, and Gil-Soo Jang. "Soundscape design for the memorial space with seaside view—focussed on the seaside park with observatory located in Ttangkkeut." *Proceedings of the 18th International Conference on Acoustics (ICA).* Kyoto, 2004.

Shultis, Christopher. "Silencing the Sounded Self: John Cage and the intentionality of Nonintention." *The Musical Quartely* 79, no. 2 (1995): 312-350.

Shultis, Christopher. "Silencing the Sounded Self: John Cage and the Intentionality of Nonintention." *The Musical Quaterly* 79, no. 2 (Summer 1995): 312-350.

Siegel, Marjorie. "More than Words: The Generative Power of Transmediation for Learning." *Canadian Journal of Education/Revue Canadienne de l'education* 20, no. 4 (Autumn 1995): 455-475.

Sierksma, Fokke. *Tibet's Terrifying Deities.* Tokyo: Charles E. Tuttle, 1966.

Skånberg, A., and E. Öhrström. "Adverse health affects in relation to urban residential soundscapes." *Journal of Sound and Vibration* 250, no. 1 (2002): 151-155.

Slawson, David A. *Secret Teachings in the Art of Japanese Gardens.* New York, NY: Kodansha, 1987.

Smith, Michael L. "Technological Determinism in American Culture." In *Does Technology Drive History: The Dilemma of Technological Determinism*, edited by Merritt Roe Smith and Leo Marx, 1-36. Boston: MIT Press, 1994.

Smith, Michael L. "Technological Determinism in American Culture." In *Does Technology Drive History: the dilemma of technological determinism*, edited by Merritt Roe Smith and Leo Marx, 2-32. Boston: MIT Press, 1994.

Southworth, Michael. "The Sonic Environment of Cities." *Environment and Behavior* 1 (1969): 49-70.

Stanley-Barker, Joan. *Japanese Art*. London: Thames and Hudson, 1984.

Stauskis, Gintaras. "Japanese Gardens outside of Japan: From the Export of Art to the Art of Export." *Town Planning and Architecture* 35, no. 3 (2011): 212-221.

Swan, Peter C. *A Concise History of Japanese Art*. Tokyo: Kodansha International, 1979.

Szalapaj, Peter. *Contemporary ARchitecture and the Digital Design Process*. Oxford: Architectural Press, 2005.

Taguchi, Yoshio. *The Architecture of Yoshio Taguchi*. New York: Harry N. Abrams, 1999.

Takei, Jirō, and Marc Peter Keane, trans. *Sakuteiki, visions of the Japanese garden*. Boston, MA: Tuttle Publishing, 2001.

Tamura, Toshi. *Korakuen-shi: history of Korakuen*. Tokyo: Tokoshoin, 1929.

Tedeschi, Arturo. *Parametric Architecture with Grasshopper, Primer Guide*. Edited by Arturo Tedeschi. Brienza: Le Penseur, 2011.

Thompson, Emily. "Noise and Noise Abatement in the Modern City." In *Sense of the City*, edited by Mirko Zardini, 191-199. Montréal: Canadian Centre for Architecture, 2006.

Todorov, Tzvetan. *Grammaire du Décaméron*. The Hague: Mouton, 1969.

Torres, Anna Maria. *Isamu Noguchi: A Study of Space*. New York: The Monacelli Press, 2000.

Treib, Mark. *A Guide to the Gardens of Kyoto*. New York: Kodansha International, 2003.

Truax, Barry. *Acoustic Communication*. Westport, CT: Ablex Publishing, 2001.

—. *The Hanbook for Acoustic Ecology*. Burnaby, B.C: Cambridge Street Publishing, 1999.

Truax, Barry, and Gary W. Barett. "Soundscape in a context of acoustic and landscape ecology." *Landscape Ecology* 26 (2011): 1201-1207.

Tschumi, Christian. *Mirei Shigemori: Modernizing the Japanese Garden*. Bereley, CA: Stone Bridge Press, 2005.

Tudor, David, and John Cage. *Indeterminacy: New Aspects of Forms in Instrumental and Electronic Music*. Smithsonian/Folkways Recordings FT 3704. LP. 1959.

Turner, M. G. "Landscape ecology: the effects of patterns on progress." *Annual Review of Ecological Systems* 20 (1989): 171-197.

United Nations. *State of World Population 2007: Unleashing the potential of urban growth*. web resource. New York, NY: UNFPA, 2007.

van Tonder, Gert. "Recovery of visual structure in illustrated Japanese gardens." *Pattern Recognition Letters* 28 (2007): 728-739.

van Tonder, Gert, and Michael Lyons. "Visual Perception in Japanese rock garden design." *Axiomathes* 15 (2005): 353-371.

Varley, Paul. *Japanese Culture.* 4th Edition. Honolulu: University of Hawaii Press, 2000.

Vogt, Frank, and Rudolf Willie. "TOSCANA-A graphical tool for analyzing and exploraing data." In *Lecture Notes in Computer Sciences,* edited by Tamassia R. and I. G. Tollis, 226-233. Heidelberg: Springer-Verlag, 1995.

Wölfflin, Heinrich. *Principles of Art History.* New York: Dover, 1932.

Walters, John Kevin. *Blobitecture: Waveform Architecture and Digital Design.* Glouster: Rockport Publishers Inc., 2003.

Westerkamp, Hildegard. "Soundwalking as Ecological Practice." In *The East Meets the West in Acoustic Ecology,* edited by Keiko Torigoe, Tadahiko Imada and Kozo Hiramatsu, 84-89. Hirosaki: Hirosaki University, 2006.

Weston, Richard. "Koshino House." In *Plans, Sections and Elevations: Key Buildings of the Twentieth Century,* by Richard Weston, 186-187. London: Laurence King Publishing, 2004.

White, Merry. *The Japanese, can they go home again?* Princeton: Princeton University Press, 1992.

Wigley, Mark. *Constant's New Babylon.* Rotterdam: Witte de With, 1998.

Wilhelm, Hellmut and Richard Wilhelm. *Understanding the I-Ching: The Wilhelm lectures on the Book of Changes.* Princeton: Princeton University Press, 1995.

Willie, Rudolph. "Restructuring lattice theory: an approach based on hierachies of concepts." Vol. 5548. In *Formal Concept Analysis,* edited by Sébastien Ferré and Sebastian Rudolph, 314-339. Berlin: Springer Verlag, 2009.

Wilson, Andrew. *Influential Gardeners: The Designers who shaped 20th-Century garden style.* London: Mitchell Beazley, 2002.

Wolff, Karl E. "A First Course in Formal Concept Analysis: how to undstand line diagrams." *SoftStat'93 Advances in Statistical Software* 4 (1994): 429-438.

Woods, Lebbus. Foreword to *Ambiguous Spaces,* by Nannette Jackowski and Ricardo de Ostos, edited by Linda Lee, 5. New York: Princeton Architectural Press, 2008.

World Health Organisation. "Guidelines for Community Noise." Edited by D. H. Schwela, B. Berglund and T. Lindvall. Geneva: WHO, 1999.

Yamaguchi, Keita, Isao Nakajima, and Masashi Kawasaki. "The Application of the Surrounding Landform to the Landscape Design in Japanese Gardens." *WSEAS Transactions on Environment and Development* 8, no. 4 (August 2008): 655-665.

Yang, Wei, and Jian Kang. "A cross-cultural study of soundscape in urban open public spaces." *Proceedings of the Tenth International Congress on Sound and Vibration.* Stockholm, 2003.

Yang, Wei, and Jian Kang. "Soundscape and Sound Preferences in Urban Squares: a Case Study in Sheffield." *Journal of Urban Design* (Routledge) 10, no. 1 (February 2005): 61-80.

Yevtushenko, Serhiy A. "Systems of data analysis 'Concept Explorer'." *Proceedings of the 7th National Conference on Artifical intelligence KII-2000*. Moscow: KII-2000. 127-134.

Young, D., M. Young, and T. Yew. *The Art of the Japanese Garden*. Tokyo: Tuttle Publishing, 2005.

Zonneveld, I. S. "Scope and concepts of landscape ecology as an emerging science." In *Changing Landscape: An Ecological Persepctive*, edited by R. T. T. forman, 3-20. New York: Springer-Verlag, 1990.